# Issues, Evidence, and You

SEPUP

Science Education for Public Understanding Program

University of California at Berkeley

Lawrence Hall of Science

This project was supported, in part,
by the
**National Science Foundation**
Opinions expressed are those of the authors
and not necessarily those of the Foundation

LAB-AIDS®, Inc. Publishing Division          Ronkonkoma, NY

Part I photo credits: pp. 40 and 43—Courtesy of East Bay Municipal Utility District; pp. 66 and 133—Courtesy of *ChemEcology*; p. 125—Drawing by Rose Craig; p. 148—Courtesy of Fain Drilling and Pump Co., Inc., Valley Center, CA.

Part II photo credits: p. 3—Illustrations of hand through evolution from *The Human Evolution Coloring Book* by Adrienne Zihlman. Copyright ©1982 by Coloring Concepts, Inc. Reprinted by permission of HarperCollins Publishers, Inc.; p. 8 (photo 2)—Courtesy of The Field Museum (Chicago, IL), Neg. #39548; pp. 8 (photos 4 and 5), 9 (photos 8, 9, and 11), 16 (upper right), 19 (photo 6), and 44 (upper left)—Phoebe A. Hearst Museum of Anthropology (University of California, Berkeley); p. 16 (upper left)—The Nelson-Atkins Museum of Art (Kansas City, MO) (Purchase: Nelson Trust) 61–15; pp. 16 (lower right and lower left), 36, 44 (right and lower left), 45, 46, 74, 76, 104, 107, and 127—Noa Shein; p. 19 (photo 5)—Courtesy of Wassim Jabi (University of Michigan); p. 42—Pioneer Press, Bill Lane, 1990; p. 73—Courtesy of The Packaging Store, Inc.; pp. 77 and 98—Courtesy of *ChemEcology*; pp. 91 and 137—Courtesy of Chemical Waste Management, Inc.; p. 135—Courtesy of Hewlett-Packard Company.

Part III photo credits: pp. 23 and 24—New York State Electric and Gas Corporation; p 23 (right)—®Reddy Kilowatt used with permission of The Reddy Corporation International, Albuquerque, NM; p. 66—Wind and biomass photographs courtesy of the Electric Power Research Institute; solar photograph courtesy of Siemens Solar Industries; hydroelectric power photograph courtesy of the U.S. Department of Energy; p. 67—Courtesy of the U.S. Department of Energy; p. 75—Courtesy of the Electric Power Research Institute; p. 83—Courtesy of the Cadillac Motor Car Division; p. 87—Courtesy of Jared Schneidman Design and *Popular Science* (June 1994); p. 93—Courtesy of the Ford Motor Company.

Part IV photo credits: p. 2—Courtesy of the Morningstar Factory, CA; p. 3 (second photo from top)—Courtesy of Hewlett-Packard Company; p. 3 (bottom photo)—Courtesy of Exxon Company, USA; p. 6—Courtesy of *ChemEcology*; p. 18—Drawing by Rose Craig; p. 27—Photo of cookies courtesy of Otis Spunkmeyer; p. 46—Reprinted courtesy of the Institute of Archaeology, University of California, Los Angeles. Photo by Georgia Lee.

SEPUP
Lawrence Hall of Science
University of California at Berkeley
Berkeley, CA 94720-5200
e-mail: sepup@uclink.berkeley.edu

©1996 The Regents of the University of California
ISBN: 1-887725-06-7

1 2 3 4 5 6 7 8 9    00 99 98 97 96

Produced by:
LAB-AIDS®, Inc. Publishing Division
17 Colt Court
Ronkonkoma, NY 11779

Distributed by:
Sargent-Welch VWR Scientific
911 Commerce Court
Buffalo Grove, IL 60089

# Acknowledgments

## SEPUP Staff

Dr. Herbert D. Thier, Program Director
Mark Koker, Teacher Associate, Assistant Director 1990–94
Dr. Robert Horvat, Program Development Center Coordinator, Associate Director June 1995–96
Dr. Barbara Nagle, Teacher Associate, Development Coordinator
Mike Reeske, Teacher Associate, Living With Chemicals Coordinator
Stephen Rutherford, Teacher Associate, 1993-94
Pamela T. Boykin, Teacher Associate, 1994–95
Dr. Kathryn Sloane-Weisbaum, Assessment Co-Director
Dr. Mark Wilson, Assessment Co-Director
Robin Henke, Assessment
Lily Roberts, Assessment
Steve Ruis, Cooperating Science Educator
Dr. Peter Kelly, Research Associate
Dr. Magda Medir, Research Associate
Dr. Marinda Li Wu, Coordinator, Private Sector Relations, 1994–95
Miriam Shein, Publications Coordinator
Londie Peterson, Program Representative
Janice Gardner-Loster, Administrative Assistant/Editor

Major contributors to the development and writing of *Issues, Evidence and You* were (listed alphabetically):
Robert Horvat
Barbara Nagle
Mike Reeske
Stephen Rutherford
Herbert D. Thier

Significant contributions were also made by Pamela T. Boykin and Mark Koker.

Teacher participants in the 1994 summer writing and editorial conferences were:
Eddie Bennett, New York City
Kathaleen Burke, Western New York
Carolyn Delia, Michigan
Cynthia Detwiler, Kentucky-Louisville Area
Richard Duquin, Western New York
Anne Little, North Carolina
Donna Markey, California-San Diego County
Susie Nally, Kentucky-Lexington Area
Lori Sheppard-Gillam, Alaska
Linda Sherrill, Oklahoma
Dorothy Trusclair, Louisiana

The following people were responsible for the production of this book:
Editing: Margo Crabtree
Copyediting: Janice Gardner-Loster
Design and Layout: Miriam Shein
Cover Art: Carol Bevilacqua
Cover Prepress: University of California Printing Services
Research: Marcelle Siegel

From the beginning of the development process, the assessment team worked with us to ensure an integrated assessment system and to keep us focused on key concepts and variables of the course framework. Major contributors to development of the assessment system were:
Robin Henke
Lily Roberts
Kathryn Sloane-Weisbaum
Chris White
Mark Wilson

Scientific review was provided by the following individuals:
Gary Arant, General Manager, Valley Center Municipal Water District
Joe Davis, retired science teacher
Rollie Myers, Chemistry Department, University of California at Berkeley
Steve Ruis, Chemistry Department, American River College

# Acknowledgments (cont.)

### Program Development Centers—Directors and Teachers

The classroom is SEPUP's laboratory for development. We are extremely appreciative of the following center leaders and teachers who taught the program during the 1993–94 and/or 1994–95 school years. These teachers, and their students, contributed significantly to improving the scope, quality, and teachability of the course.

Alaska Center, Donna York
    Kim Bunselmeyer, Linda Churchill, James Cunningham, Patty Dietderich, Lori Gillam, Gina Ireland-Kelly, Mary Klopfer, Jim Petrash, Amy Spargo

California-San Bernardino County Center, Dr. Herbert Brunkhorst, 1993–94
    William Cross, Alan Jolliff, Kimberly Michael, Chuck Schindler

California-San Diego County Center, Mike Reeske and Marilyn Stevens (Co-Directors)
    Pete Brehm, Donna Markey, Susan Mills, Barney Preston, Samantha Swann

California-San Francisco Area Center, Stephen Rutherford
    Michael Delnista, Cindy Donley, Judith Donovan, Roger Hansen, Judi Hazen, Catherine Heck, Mary Beth Hodge, Mary Hoglund, Mary Pat Horn, Paul Hynds, Margaret Kennedy, Carol Mortensen, Bob Rosenfeld, Jan Vespi

Colorado Center, John E. Sepich
    Mary Ann Hart, Lisa Joss, Geree Pepping-Dremel, Tracy Schuster, Dan Stebbins, Terry Strahm

Connecticut Center, Dave Lopath
    Harald Bender, Laura Boehm, Antonella Bona-Gallo, Joseph Bosco, Timothy Dillon, Victoria Duers, Valerie Hoye, Bob Segal, Stephen Weinberg

Kentucky-Lexington Area Center, Dr. Stephen Henderson and Susie Nally (Co-Directors)
    Stephen Dilly, Ralph McKee II, Barry Welty, Laura Wright

Kentucky-Louisville Area Center, Ken Rosenbaum
    Ella Barrickman, Pamela T. Boykin, Bernis Crawford, Cynthia Detwiler, Denise Finley, Ellen Skomsky

Louisiana Center, Dr. Sheila Pirkle
    Kathy McWaters, Lori Ann Otts, Robert Pfrimmer, Eileen Shieber, Mary Ann Smith, Allen (Bob) Toups, Dorothy Trusclair

Michigan Center, Phillip Larsen, Dawn Pickard, and Peter Vunovich (Co-Directors), 1993–94
    Ann Aho, Carolyn Delia, Connie Duncan, Kathy Grosso, Stanley Guzy, Kevin Kruger, Tommy Ragonese

New York City Center, Arthur Camins
    Eddie Bennett, Steve Chambers, Sheila Cooper, Sally Dyson

North Carolina Center, Dr. Stan Hill and Dick Shaw (Co-Directors)
    Kevin Barnard, Ellen Dorsett, Cameron Holbrook, Anne M. Little

Oklahoma Center, Shelley Fisher
    Jill Anderson, Nancy Bauman, Larry Joe Bradford, Mike Bynum, James Granger, Brian Lomenick, Belva Nichols, Linda Sherrill, Keith Symcox, David Watson

Pennsylvania Center, Dr. John Agar
    Charles Brendley, Gregory France, John Frederick, Alana Gazetski, Gill Godwin

Washington, D.C. Center, Frances Brock and Alma Miller (Co-Directors)
    Vasanti Alsi, Yvonne Brannum, Walter Bryant, Shirley DeLaney, Sandra Jenkins, Joe Price, John Spearman

Western New York Center, Dr. Robert Horvat and Dr. Joyce Swartney (Co-Directors)
    Rich Bleyle, Kathaleen Burke, Al Crato, Richard Duquin, Lillian Gondree, Ray Greene, Richard Leggio, David McClatchey, James Morgan, Susan Wade

# Issues, Evidence and You

## Table of Contents

*Please note that each part of this book is numbered separately and may be identified by its color.*

### Part 1 Water Usage and Safety

Introduction

#### Section A: Questions About a Glass of Water
- Overview .................................................................................................................. 2
1. Drinking-Water Quality .......................................................................................... 3
2. Exploring Sensory Thresholds ............................................................................... 8
3. Concentration ........................................................................................................ 14
4. Mapping Death ..................................................................................................... 21
5. John Snow and the Continued Search for Evidence ............................................. 26
6. Detecting Contaminated Water: The Biological Risks .......................................... 32
7. Chlorination: How Much Is Just Enough? ............................................................. 40

#### Section B: Threshold, Toxicity, and Risk
- Overview .................................................................................................................. 49
8. Chicken Little, Chicken Big: A Threshold Story .................................................... 50
9. Lethal Toxicity ....................................................................................................... 55
10. Risk Comparison ................................................................................................... 64
11. The Injection Problem ........................................................................................... 68
12. The Peru Story ....................................................................................................... 72

#### Section C: Chemical Testing
- Overview .................................................................................................................. 79
13. Chemical Detective ............................................................................................... 80
14. Acids, Bases, and Indicators ................................................................................. 84
15. Serial Dilutions of Acids and Bases ...................................................................... 90
16. Acid-Base Neutralization ...................................................................................... 96
17. Quantitative Analysis of Acid .............................................................................. 102
18. The "Used" Water Problem ................................................................................. 105
19. Is Neutralization the Solution to Pollution? ........................................................ 108
20. Water Quality ....................................................................................................... 112

#### Section D: Investigating Groundwater
- Overview ................................................................................................................ 120
21. Trouble in Silver Oaks ......................................................................................... 121
22. Water Movement Through Earth Materials ........................................................ 129
23. The "Ins and Outs" of Groundwater .................................................................... 132
24. Investigating Contamination Plumes .................................................................. 142
25. Testing the Waters ............................................................................................... 148
26. Mapping It Out ..................................................................................................... 150
27. Practicing Presentation Skills ............................................................................. 155
28. Cleaning It Up ...................................................................................................... 157

# Table of Contents (cont.)

## Part 2  Materials Science
    Introduction ............................................................................................................. 1
29. Differentiating Humans from Other Organisms ................................................. 3
30. Materials Through Time ....................................................................................... 8
31. Properties of Materials ........................................................................................ 17
32. Conductors and Insulators ................................................................................. 24
33. Preventing Corrosion .......................................................................................... 29
34. Bag It! Paper or Plastic? ..................................................................................... 42
35. Properties of Plastics .......................................................................................... 46
36. Synthesizing Polymers ........................................................................................ 57
37. Paper Clip Polymers ............................................................................................ 63
38. Which Packing for Mike's Games? .................................................................... 73
39. Comparing Garbage ............................................................................................ 79
40. Investigating Sanitary Landfills ......................................................................... 82
41. Products of Hazardous Waste Incineration ...................................................... 91
42. Recycling Materials ............................................................................................. 98
43. Chemical Change: The Aluminum-Copper Chloride Reaction ....................... 107
44. Reducing and Recycling Hazardous Materials ................................................ 118
45. Source Reduction ............................................................................................... 123
46. Integrated Waste Management ......................................................................... 137

## Part 3  Energy
    Introduction ............................................................................................................. 1
47. Batteries: Energy and Disposal ........................................................................... 2
48. Electroplating ....................................................................................................... 15
49. Electrical Appliance Survey ................................................................................ 19
50. Investigating Energy Transfer ............................................................................ 27
51. Developing an Energy Savings Plan .................................................................. 32
52. Ice and Energy Transfer ..................................................................................... 37
53. Quantifying Energy: Calorimetry ....................................................................... 45
54. Efficiency: Energy Changes and Waste ............................................................. 56
55. Electrical Energy: Sources and Transmission .................................................. 65
56. Energy from the Sun ........................................................................................... 69
57. Controlling Radiant Energy Transfer ................................................................ 78
58. Designing an Energy-Efficient Car .................................................................... 80

## Part 4  Environmental Impact
    Introduction ............................................................................................................. 1
59. Industry Comes to Town ...................................................................................... 2
60. Pinniped Island .................................................................................................... 18
61. Synthesis, Testing, and Redesign of White Glue .............................................. 22
62. Scaling Up Glue Production ............................................................................... 27
63. Packaging and Labeling Products ..................................................................... 30
64. Planning a Factory .............................................................................................. 32
65. An Island Parable ................................................................................................ 46

# Part One

# Water Usage and Safety

# Part 1

# Table of Contents

*Introduction*

*Section A: Questions About a Glass of Water*

    Overview ................................................................................................2
1. Drinking-Water Quality .......................................................................3
2. Exploring Sensory Thresholds ...........................................................8
3. Concentration ....................................................................................14
4. Mapping Death ..................................................................................21
5. John Snow and the Continued Search for Evidence ....................26
6. Detecting Contaminated Water: The Biological Risks .................32
7. Chlorination: How Much Is Just Enough? ......................................40

*Section B: Threshold, Toxicity, and Risk*

    Overview ..............................................................................................49
8. Chicken Little, Chicken Big: A Threshold Story ............................50
9. Lethal Toxicity ....................................................................................55
10. Risk Comparison ...............................................................................64
11. The Injection Problem ......................................................................68
12. The Peru Story ...................................................................................72

*Section C: Chemical Testing*

    Overview ..............................................................................................79
13. Chemical Detective ...........................................................................80
14. Acids, Bases, and Indicators ............................................................84
15. Serial Dilutions of Acids and Bases ................................................90
16. Acid-Base Neutralization ..................................................................96
17. Quantitative Analysis of Acid .........................................................102
18. The "Used" Water Problem ............................................................105
19. Is Neutralization the Solution to Pollution? ................................108
20. Water Quality ...................................................................................112

*Section D: Investigating Groundwater*

    Overview ............................................................................................120
21. Trouble in Silver Oaks ....................................................................121
22. Water Movement Through Earth Materials .................................129
23. The "Ins and Outs" of Groundwater ..............................................132
24. Investigating Contamination Plumes ...........................................142
25. Testing the Waters ..........................................................................148
26. Mapping It Out ................................................................................150
27. Practicing Presentation Skills ........................................................155
28. Cleaning It Up ..................................................................................157

# Section A: Questions About a Glass of Water
## Activities 1–7

### Overview

Water is the essential medium within which life functions. This is as true of plants and animals that live on land as it is of those that live in water. One of the most important properties of water is its ability to dissolve many other substances. Our blood, which is mostly water, carries the oxygen we breathe and the dissolved nutrients from the food we eat to every cell in our bodies, and then carries out the dissolved waste products.

Because of the important role that water plays in our lives, the first seven activities of the SEPUP course explore the topic of drinking water quality and the primary health risks associated with poor water quality. How do you know that the water you drink is pure or safe? Can you detect contaminated water? What is done to the drinking water in your community to ensure its safety? Whose responsibility is it?

As you investigate these questions and discuss the evidence you have gathered, you will be learning and practicing skills that will be used for the rest of the SEPUP course.

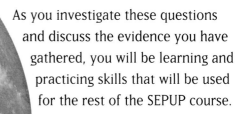

# Activity 1
# Drinking-Water Quality

## Introduction

### Can You Taste the Difference?

You have probably tried bottled spring water. Every year in the United States, people spend over $2.7 billion on bottled water. Do you think people can tell the difference between tap water, distilled water (water without dissolved solids), and bottled spring water?

## Challenge

Your challenge is to determine if you or your classmates can identify the more expensive bottled spring water in a taste test.

Activity 1    3

# 1

## Drinking-Water Quality

### Materials

*For each group of four students:*

- Three samples of water labeled A, B, and C
- Four small tasting cups—one for each student

### Procedure

**Safety**

Use your own tasting cup. Reminder: Only taste chemicals in the lab when instructed to do so by your teacher.

1. Set up your investigation report according to your teacher's instructions.

2. Fill each person's tasting cup half full of water sample A.

3. Observe the appearance of the water sample. Smell the sample. Finally, taste the sample. Drink all of the water in the cup.

4. Record your observations of appearance, smell, and taste in your Journal. Use a larger version of a data table like the sample below. Do not share your results with other members of the group.

*Water Tasting (reproduce in your Journal)*

| Water Sample | Observations | | | Proposed Sample Identity |
|---|---|---|---|---|
| | Appearance | Smell | Taste | |
| A | | | | |
| B | | | | |
| C | | | | |

5. Repeat steps 2 through 4 for water samples B and C.

6. Identify the sample that you think is the bottled spring water. Record your result in your data table.

Activity 1

*Drinking-Water Quality*

7. When everyone in the group is finished, have each person share his or her results with the group. Everyone should give a reason for the sample identified as bottled water.

8. The group should now try to reach consensus about the identity of the spring water. Record the result in a data table like the one below.

---

**GROUP CHOICE** for identity of bottled spring water _____

**REASON** _____

_____

_____

**CLASS CHOICE** for identity of bottled spring water _____

**REASON** _____

_____

_____

**ACTUAL SAMPLE IDENTITY**

   SAMPLE A _____

   SAMPLE B _____

   SAMPLE C _____

---

 **Data Processing**

Record in your Journal the answers to these questions. Follow your teacher's policy on the format for your answers.

1. Was the spring water the sample that tasted best to you?

2. Would you spend the extra money on spring water after the taste test? Why or why not?

3. What other information about bottled, tap, and distilled water would you like to know before you decide which water to drink?

# Drinking-Water Quality

## Introduction

### Purity in a Bottle

Do you think bottled spring water is worth the extra money? Is there more information you need to know? Sometimes bottled spring water may not be what you think it is.

## Challenge

> Your challenge is to read this article and determine if it is a fair treatment of the issue: "Should I drink bottled water?"

Millions of Americans choose to pay for bottled water rather than drink their local tap water. But some brands of bottled water may not contain the pristine spring water described on their labels! Many states have begun to require bottled-water producers to disclose the source of their water and to meet specific standards for water quality. When officials in North Carolina discovered that the source for eight popular brands of bottled water was underground wells and that the water was pumped to the surface from holes drilled into these wells, the bottled water was removed from store shelves. Although spring water may or may not be purer than well water (both come from groundwater), North Carolina insists that only water from naturally flowing springs may be labeled "spring water."

This case points out the need for accurate labeling so consumers can judge whether bottled water may offer them any benefits over drinking the local tap water—especially given the prices charged for some of the brands. Is bottled water really an untapped source from paradise, as some advertisements suggest, or is it only high-priced tap water? The U.S. House

## Drinking-Water Quality

Energy and Commerce Committee delivered a setback to the bottled-water industry when it announced the results of a 1991 study that found that 25% of high-priced bottled waters came from the same source as the local tap water, and another 25% of the bottlers were not able to document the source of their water at all!

Are bottled waters worth the extra $2.7 billion a year consumers spend on them? In the same House Energy and Commerce Committee study, 31% of bottled water exceeded the federally allowable levels of microbiological contamination. In addition, most states do not require labeling for such dangerous contaminants as nitrates and lead, whereas tap water, by federal law, is constantly checked for these and nearly 200 other chemicals. In one of our largest states, the law permits bottlers to label their product "spring water"—for which consumers pay a premium price—if as little as 10% of the water actually comes from a spring. The rest can come straight from the tap. Can bottlers respond to the concerns and regain public confidence? With bottled-water sales dropping to an annual 3% rate of growth, down from 500% growth in the 1980s, they may have a long way to go.

### Questions

1. What is the point of view expressed in the magazine article?
2. What *evidence* is included in the article? (Evidence generally refers to facts or observations that can be tested or checked.)
3. Do you agree or disagree with the article? Cite *evidence* to support your position.
4. What additional evidence would you like before deciding which water to drink?

# Activity 2

## Exploring Sensory Thresholds

### Introduction

**Thresholds of Vision, Taste, and Smell**

In the last activity you tried to identify bottled spring water by taste. In this activity, you will investigate the ability of two more senses—sight and smell—to detect substances dissolved in water.

### Challenge

> Determine which one of your senses—taste, smell, or sight—can detect the lowest concentration of a drink mix solution. Use a bar graph to compare the range of these senses for all the members of your class.

Can our senses tell us if the water in this river is safe to drink?

*Exploring Sensory Thresholds*

## Materials

*For each group of four students:*

- Ten 9-ounce plastic cups, labeled 1–10
- Four small tasting cups
- One paper towel
- One 2-liter bottle of tap water, labeled "Tap Water"
- One stirring stick

*For each student:*

- Graph paper (or photocopy of Transparency Master 2)

## Procedure

### Part One

1. Set up your investigation report.

2. Based on your own personal experiences, predict which of your senses—sight, smell, or taste—can detect the lowest concentration of drink mix dissolved in water. Record your prediction. In your Journal, give two reasons for your choice.

*Continued on next page* →

# Exploring Sensory Thresholds

**Procedure (continued)**

**Safety**

Use clean plastic cups for this investigation. Do not share the tasting cups, and do not add anything to the plastic cups, including fingers! Reminder: Only taste chemicals in science class when instructed to do so by your teacher.

3. Your teacher will fill Cup 1 half full of concentrated drink mix solution.

4. Fill the cup to the 200-mL mark with tap water from the bottle. Mix with the stirring stick.

5. Pour half of the liquid from Cup 1 into Cup 2 so that the levels in the two cups are equal.

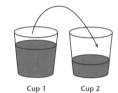

6. Fill Cup 2 to the 200-mL mark with tap water from the bottle. Mix well.

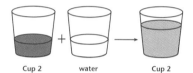

7. Now pour half of the liquid from Cup 2 into Cup 3 so that the levels in the two cups are equal.

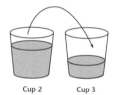

8. Fill Cup 3 with water to the 200-mL line with tap water. Mix.

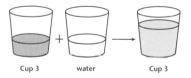

9. Repeat this procedure, using Cups 4 through 9, until all nine cups have been filled. Note: Cup 9 will be completely full of solution.

10. To Cup 10 add 200 mL of tap water.

11. Record the dilutions in a data table like the one on page 12.

# Exploring Sensory Thresholds

*Part Two*

1. Each person in your group should take a clean small tasting cup. This will be your tasting cup, so do not mix it up with the others. Beginning with the tap water in Cup 10, pour a small amount of water (about 15 mL or enough to half fill the tasting cup) into your small tasting cup. This will represent the control—a solution that contains no drink mix. Look at the sample, smell it, and then take a taste. Record your observations. Empty your cup after each taste.

2. Pour a small amount from Cup 9 into each person's tasting cup.

3. Again, look at the sample, smell it, and then take a taste of the solution. Do not tell your group whether you can see, taste, or smell anything. Record your observations in your data table.

4. Repeat the process for Cups 8 through 1, in that order. For each cup, record whether you are able to see, smell, and taste the drink mix in each cup. Results of the tests should be kept private until all 10 solutions have been tested. Take care not to alert your partners through body language or facial expressions whether you can detect the drink mix.

5. Circle on your data table each instance where you were first able to detect the drink mix (for example, circle the box under "Appearance" where you were able to see the mix).

6. Record, on the threshold transparency, the cups in which you were first able to see, smell, and taste the drink mix. Remember, you started with the very dilute drink mix in Cup 9, so the first cup in which you detected the drink mix was the highest, not the lowest, cup number.

7. Share your results with the other members of your group of four.

8. Complete the Data Processing section.

9. Clean up as directed by your teacher.

# Exploring Sensory Thresholds

*Sensory Thresholds*

| Cup | Dilution | Appearance? | Smell? | Taste? |
|---|---|---|---|---|
| 10 | | | | |
| 9 | | | | |
| 8 | | | | |
| 7 | | | | |
| 6 | | | | |
| 5 | | | | |
| 4 | | | | |
| 3 | | | | |
| 2 | 1/2 | | | |
| 1 | 1 | | | |

*Exploring Sensory Thresholds*

### Data Processing

1. Use the bar graph sheet provided by your teacher or a sheet of graph paper to make a bar graph for each of the three senses. Record the information from the class results on your graph. Use a different color for each test.

2. According to the graph, in what cup were the most students first able to:

   a. Detect the color of the drink mix?

   b. Detect the smell of the drink mix?

   c. Detect the taste of the drink mix?

   These represent the most common values (modes) for vision, taste, and smell thresholds.

If everyone first saw the drink mix in Cup 5, your bar graph would look like this:

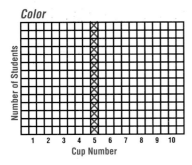

3. From our investigation, which sense is best able to detect the drink mix at low concentrations?

4. How did your personal results compare to those of the other members of your group?

5. Give two reasons that could explain why different people reported first seeing, tasting, or smelling the drink mix in different cups.

6. Describe an experiment you could do to check one reason you suggested for people sensing the drink mix at different concentrations. Include a statement of your problem and how you would do the experiment.

7. Based on the class discussion:

   a. Define the term *threshold*.

   b. Give an example of a threshold from this activity.

   c. Give an example of a threshold for a substance you use.

   d. Represent the concept of threshold. Use a labeled drawing and written description to express your ideas.

Activity 2

# Activity 3

## Concentration

### Introduction

**Parts per Million**

In the last activity you found that there were concentrations of powdered drink mix that you couldn't see, smell, or taste; but you didn't know exactly what the concentrations were. In this activity you will find a way to describe the amount of food coloring in a solution if the amount is very, very small.

### Challenge

Your challenge is to use parts per million to describe the concentration of a solution.

We dilute orange juice concentrate with water before we drink it.

# Concentration

## Materials

*For each group of four students:*

- One 30-mL dropping bottle of red food coloring (10% solution)

*For each pair of students:*

- One SEPUP tray
- One dropper
- One 30-mL dropping bottle of water
- Student Sheet 3, "Serial Dilution Template"

# Concentration

## Procedure

1. Place the Serial Dilution Template (Student Sheet 3) under the SEPUP tray.

2. Put 10 drops of 10% red food coloring into small Cup 1 and put one drop into small Cup 2.

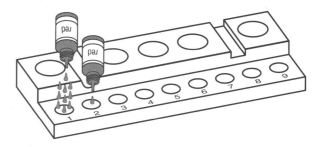

3. To small Cup 2, add 9 drops of water. Mix the solution by drawing it up into the dropper. Then gently squeeze the bulb until the dropper is empty, carefully putting the liquid back into Cup 2.

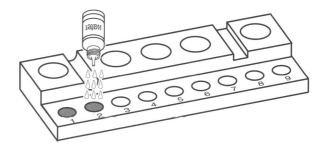

4. Using the dropper, transfer one drop of the solution in Cup 2 to Cup 3. Return any excess to Cup 2.

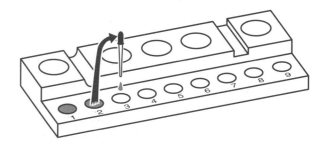

# Concentration

5. Add 9 drops of water to Cup 3. Use the dropper to mix the solution in Cup 3, and transfer one drop to Cup 4. Return any excess to Cup 3.

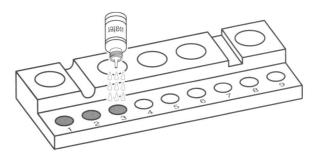

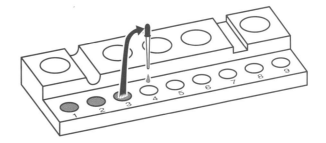

6. Add 9 drops of water to Cup 4. Mix. Transfer one drop to Cup 5. Add 9 drops of water to Cup 5. Mix.

7. Continue the process through Cup 9, each time taking a drop of the solution from the previous cup and adding 9 drops of water.

8. Record the color of the solution in each cup in a data table in your Journal. A sample data table is shown on page 18.

9. Determine the concentration of the solution for each cup as a part of food coloring per amount of solution, and record it in your data table.

10. Answer the Data Processing questions.

# 3

## Concentration

*Serial Dilution*

| Cup | Color | Concentration (parts of dye per parts of solution) ||
|---|---|---|---|
| | | parts per ____ | % |
| 1 | | 1 part in 10 | 10% |
| 2 | | 1 part in ____ | |
| 3 | | 1 part in ____ | |
| 4 | | | |
| 5 | | | |
| 6 | | | |
| 7 | | | |
| 8 | | | |
| 9 | | | |

Activity 3

# Concentration

 **Data Processing**

1. Which is more dilute, Cup 1 or Cup 2? How do you know this?

2. If Cup 1 has a concentration of one part in 10, and Cup 2 has 1/10 the concentration of Cup 1, what is the concentration of Cup 2?

3. Which cup has a concentration of one part per million?

4. What is the number of the cup in which the solution first appeared colorless? What is the concentration in parts of dye per parts of solution in this cup? (Express the answer for concentration as one part per ____.)

5. What are the possible reasons for student differences in reporting the cup in which the solution first appeared colorless? (Hint: Consider the idea of threshold.)

6. Do you think that any of the food coloring is present in this cup of diluted solution even though it appears colorless? Explain.

7. Explain how you could do an experiment to prove that there is actually some red food coloring in this cup.

8. How would you explain what a million is to a young child?

9. If you change the solution of food coloring in Cup 1 from one part in 10 (10%) to five parts in 10 (50%), what would the concentration of the food coloring be in Cup 6?

*Activity 3*

# 3

## Concentration

### Introduction

**Some Interesting Comparisons**

Now that you've spent some time trying to understand or picture just what one in a million (or even one in a billion or one in a trillion) means, here is a short list of comparisons.

### Challenge

Can you think of a one in a million comparison yourself?

- One part per million is one second in 12 days of your life.
- One part per billion is one second in 32 years of your life.
- One part per million is one penny out of $10,000.
- One part per billion is one penny out of $10,000,000.
- One part per million is one pinch of salt on 20 pounds of potato chips.
- One part per billion is one pinch of salt on 10 tons of potato chips.
- One part per million is one inch out of a journey of 16 miles.
- One part per billion is one inch out of a journey of 16,000 miles.
- One part per million is approximately one bad apple in 2,000 barrels.
- One part per billion is approximately one bad apple in 2,000,000 barrels.
- One part per billion is one square foot in 36 square miles.
- One part per trillion is a postage stamp in an area the size of Texas.

Activity 3

# Activity 4

## Mapping Death

### Introduction

**The Cholera Story**

Picture yourself in London in 1832. What do you think life would be like? How would you dress? What kind of food would you eat? What would the air be like? What kind of house would you live in?

### Challenge

Your challenge is to place yourself in London in 1832 and imagine what it would be like if a member of your family were struck with cholera.

Today a tavern named in honor of John Snow stands in the area of London where the cholera epidemic took place.

# 4 Mapping Death

The following account describes Dr. O'Shaughnessy's observations of a cholera victim in 1832:

> Wanting to acquaint myself with the celebrated cholera, I traveled down to (London) from Edinburgh, prepared yet unprepared, dear sirs. I saw a face, a girl I never can forget, even were I to live beyond man's natural age.
>
> The girl lay...in a low-ceilinged room. I bent to examine her. The color of her skin—a silver blue, lead colored, ghastly tint; eyes sunk deep into deep sockets as though driven back or counter-sunk like nails, her eyelids black, mouth squared as if to bracket death; fingers bent, inky in their hue. Pulse all but gone at the wrist.
>
> It (is) not easy for survivors to forget a cholera epidemic.... The onset of cholera is marked by diarrhea, acute spasmodic vomiting, and painful cramps. Consequent dehydration (the victim can lose up to 5 gallons of liquid in 24 hours), often accompanied by cyanosis [the body turns blue], gives the sufferer a characteristic and disquieting appearance: his face blue and pinched, his extremities cold and darkened, the skin of his hands and feet drawn and puckered....Death may intervene within a day, sometimes within a few hours of the appearance of the first symptoms. And these symptoms appear with little or no warning.
>
> (*From* The Cholera Years *by C. E. Rosenberg, 1962*)

# Mapping Death

## Introduction

### Cholera Deaths

In 1849, another outbreak of cholera killed over 500 people—rich and poor, young and old—in South London. John Snow, a medical doctor in England, thought that by checking the city's death records and mapping exactly where people were living when they died, he might find some clues about what was causing the disease.

## Challenge

Your challenge is to examine the deaths from cholera in London in 1849 and to plot their location on the map to see if there is a pattern that could explain how the disease spreads.

## Procedure

1. With your partner, use the following listing of cholera deaths to plot the locations of the victims' homes on the London street map that your teacher provides. (You'll need to tape the two pieces together to make one larger map.)

2. Use a colored marker to put a small dot at the approximate address for each death.

3. If there is more than one death at the same location, put the other dots as close as possible to each other. The grid location number will help you find the street addresses.

# Mapping Death

*Deaths from Cholera in London in 1849*

| Date | Name | Age | Sex | Occupation | Address | Grid |
|---|---|---|---|---|---|---|
| 13 Feb | Anne Kelly | 3 | F | child | 156 Broad St., between Marshall & Little Windmill Streets | E-5 |
| 23 Feb | Edwin Drummond | 48 | M | steeplejack | 54 Little Windmill St., between Broad & Silver Streets | E-5 |
| 18 Mar | Patty Orford | 23 | F | seamstress | 160 Broad St., near corner of Little Windmill Street | E-5 |
| 20 Mar | Sue Burton | 22 | F | seamstress | 16 Queen St., near the corner of Little Windmill Street | H-3 |
| 27 Mar | Patrick Kelly | 39 | M | banker | 156 Broad St., between Marshall & Little Windmill Streets | E-5 |
| 28 Mar | John Kelly | 8 | M | child | 156 Broad St., between Marshall & Little Windmill Streets | E-5 |
| 3 April | Mary Thornley | 45 | F | governess | 300 Marshall St., between Broad & Silver Streets | E-6 |
| 9 April | Thomas Topham, Jr. | 19 | M | butcher | 8 New St., across from the brewery | E-4 |
| 9 April | William O'Toole | 41 | M | indigent | Poland Street Work House | D-6 |
| 13 April | Margaret Kelly | 37 | F | housewife | 156 Broad St., between Marshall & Little Windmill Streets | E-5 |
| 21 April | Richard Raleigh | 13 | M | student | 173 Broad St., between Poland & Marshall Streets | D-5 |
| 24 April | Katherine Nelson | 1 | F | child | 426 Wardour St., next to the Brewery Yard | D-3 |
| 25 April | Russ Rufer | 30 | M | steeplejack | 54 Little Windmill St., between Broad & Silver Streets | E-5 |
| 29 April | Sarah Kelly | 3 | F | child | 156 Broad St., between Marshall & Little Windmill Streets | E-5 |
| 1 May | Sir John Page | 55 | M | magistrate | 255 Broad St., between Berwick & Poland Streets | D-4 |
| 2 May | Ann Nelson | 19 | F | housewife | 426 Wardour St., next to the Brewery Yard | D-3 |
| 3 May | Agatha Summerhill | 26 | F | writer | 174 Broad St., between New and Little Windmill Streets | E-5 |
| 11 May | Barney Brownbill | 31 | M | indigent | Poland Street Work House | C-5 |
| 11 May | Rose Thornley | 53 | F | maid | 300 Marshall St., between Broad & Silver Streets | E-6 |
| 17 May | Winnifred Topham | 17 | F | factory worker | 2 Peter St., at the end | F-4 |
| 21 May | Thomas Topham | 38 | M | butcher | 2 Peter St., at the end | F-4 |
| 22 May | Winston Page | 49 | M | doctor | 1000 Regent St., near the corner of Hanover Street | D-9 |
| 27 May | Neville West | 6 | M | child | 19 Golden Square | G-6 |
| 27 May | Beatrice Braxley | 23 | F | housewife | 253 Broad St., between Berwick & Poland Streets | D-4 |
| 27 May | Eleanor Raleigh | 12 | F | student | 173 Broad St., between Poland & Marshall Streets | D-5 |

## Mapping Death

**Data Processing**

1. Describe what you see in terms of the location of the deaths. Are they scattered throughout the city, or are they bunched in a particular area?

2. Are there any clues about the cause of the disease?

3. Based on the evidence of deaths shown on the map, state two or three hypotheses or reasons that might explain how the disease is spread. (A *hypothesis* is an idea about how or why something happens.)

*London Street Map*

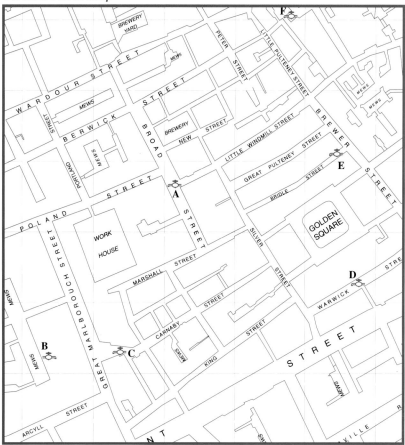

# Activity 5: John Snow and the Continued Search for Evidence

## Introduction

### Snow's Theory

Mapping the deaths due to cholera led Dr. Snow to propose that cholera was being carried in the water.

## Challenge

Think about what kinds of evidence would help Dr. Snow prove his theory on how cholera spreads.

John Snow (1813–1858)

## John Snow and the Continued Search for Evidence

Cholera causes severe diarrhea and vomiting. In India, cholera has been a health problem since 400 B.C., but it was not well known outside the Far East before 1800. In 1819, however, there was an epidemic in Europe and North America. Gradually, cholera disappeared—only to reappear suddenly in the same or a different place. Death might result within a day, sometimes within a few hours of the appearance of the first symptoms. Even more confusing, only some people would get cholera, while others living nearby would not get sick. Bad air or piled-up trash were often considered to be the cause of cholera.

In 1849, a terrible outbreak of the disease killed over 500 people who lived within a few blocks of each other in London. John Snow, a medical doctor in England, drew a detailed map of the area, showing where each victim lived. Dr. Snow discovered that the deaths came mostly from houses located near a certain public water pump. Lots of people liked to drink from this pump because of the taste and clearness of its water. Some of the deaths were reported in houses farther away and did not immediately fit the pattern.

After studying all of his data, Dr. Snow suggested that cholera was spread in the water supply by invisible bits of human waste from cholera victims.

Dr. Snow was concerned that his mapping of cholera cases in 1849 was not sufficient evidence to prove his theory concerning cholera in the water supply. Dr. Snow tested his theory by carefully reviewing the public records for 20 years before the cholera outbreak. He discovered that since 1830, about 300,000 people in one area of South London were served by just two water companies, the Lambeth Water Company and the Southwark and Vauxhall (S & V) Water Company. Originally both companies got

# 5 John Snow and the Continued Search for Evidence

This cartoon appeared in London after the cholera outbreaks.

their water from the Thames River in London. In 1840, the Lambeth Water Company changed its source of water from the Thames to a place 10 miles upstream. The other company, S & V, continued to get its water from inside London.

When cholera struck in 1854, Snow reviewed the records for the area served by the two water companies. He asked his friend and colleague Dr. John Joseph Whiting to help him gather additional evidence concerning the water sources for the city of London. He wanted to provide more proof for his theory.

# John Snow and the Continued Search for Evidence

## Introduction

### Searching for More Evidence

Dr. Snow and Dr. Whiting were two investigators who shared their data through writing. The information that they had painstakingly collected required careful organization before they could infer the cause of cholera. The letters that follow are derived from evidence of their correspondence at the time.

## Group Challenge

As a group, decide if Dr. Snow's theory about the cause of cholera is consistent with the evidence provided in the letters between Dr. Snow and Dr. Whiting.

## Individual Challenge

Make believe you are Dr. John Snow. Write a letter sharing your findings to the London Health Department. Include in your letter:
- what you think is the cause of the spread of cholera
- all the data collected in an organized form
- reasons the Health Department should believe your evidence
- what action you suggest the Health Department should take

Activity 5

# 5 John Snow and the Continued Search for Evidence

17 August 1854

John Joseph Whiting
47 Waterloo Road
London, England

Dear Dr. Whiting,

As you are aware, the dreaded cholera has once again revisited our city. I have a great interest in showing the powerful influence which invisible bits of human waste in the drinking water have on the spread of cholera. The recent outbreak has provided me with the opportunity to test my theories on the grandest scale. The S & V Water Company will not give me data so it will be necessary to collect it by going to each house. I am desirous of making the investigation myself, but I feel that I am in need of your assistance in this experiment. My current research includes the following results:

- The area of the present outbreak is served by two water companies, S & V and Lambeth.

- Each of the companies supplies both rich and poor; large and small houses; people of both sexes, all ages, all occupations, and every rank and station.

- Several years ago, the Lambeth Water Company moved its water intake pipe upstream from London's sewage-infested water. I am investigating whether this move is related to a decrease in the number of cholera cases. If so, it would support my idea that cholera is spread by human waste in the water.

To provide evidence for the support of my theories, I need to learn the water supply of each individual house where the fatal attack of cholera occurred. Would you please collect the numbers of deaths that occurred at houses supplied by the S & V Water Company?

If you are agreeable, I will send you the numbers that I have already collected from houses served by the Lambeth Water Company.

Sincerely, your friend,

John Snow, M.D.

# John Snow and the Continued Search for Evidence

30 April 1855

Dear John,

The inquiries that you requested of me were carried out with a good deal of trouble. Many hours were needed to collect the data as I had to go door to door.

The analysis of my inquiries should leave no doubt as to the correctness of your hypothesis concerning the progress of cholera. I have combined our data and report it to you as follows:

- S & V Water Company supplied 40,046 houses in which there were 1,263 deaths.
- Lambeth Water Company supplied 26,107 houses in which there were 98 deaths.
- The rest of London is served by other water companies. These companies supplied 256,423 houses in which there were 1,422 deaths.

It has been a pleasure for me to be able to assist you in this experiment of great importance to the community of London.

Sincerely, your friend,

John Joseph Whiting

# Activity 6

# Detecting Contaminated Water: The Biological Risks

## Introduction

### Microorganisms

It is hard to believe, but most living things on earth are smaller than the head of a pin! To see some of them you would need a microscope that magnifies objects at least 80–100 times larger than they normally appear. Microorganisms, which include bacteria, protozoa, viruses, yeasts, and algae, are extremely small. How small? A teaspoon of sugar weighs more than 30 trillion (30,000,000,000,000) bacteria.

## Challenge

Your challenge is to read the following article and think about how your life might be different if there were no microoganisms on Earth.

### Did you know?

- Protozoa were first seen in 1676 by Anton van Leeuwenhoek, a Dutch scientist, who made the first single-lens microscopes.

- In 1829, the name *bacteria* was coined from the Greek word *bacterion*, which means "rodlike," a common shape of many bacteria.

- The pyramids of Egypt are cemented together with a ground-up stone that is largely composed of the shells of certain protozoa.

# Detecting Contaminated Water: The Biological Risks

Microorganisms are among the oldest forms of life. In fact, the word *protozoa* means "first animal." The oldest fossils of bacterial cells, found in rocks in Africa, are over 3 billion years old. Bacteria and other microorganisms are found everywhere—in the almost airless reaches of the upper atmosphere, at the bottom of the ocean, in frozen soil, attached to rocks in hot springs, and in our large intestines (they help digest the food we eat).

Most microorganisms are harmless to humans, and many are very important for life. Bacteria and other microorganisms enrich the soil by changing the nitrogen in the air into nitrates—nutrients needed for plant growth. Bacteria also enrich the soil by acting as decomposers. Bacteria are important industrially in the production of cheese, yogurt, buttermilk, vinegar, and sauerkraut. Bacteria and other microorganisms are important in the preparation of antibiotics such as streptomycin, which is made by bacteria, and penicillin, which is made from a mold. Microorganisms help sewage disposal plants break down organic wastes. Cattle, sheep, and goats live on grass—but without bacteria and other microorganisms they would not be able to digest the cellulose in grass.

Although most bacteria and other microorganisms do not harm humans, a few are quite dangerous. Bacteria can cause severe illness in humans who eat contaminated food. In the spring of 1993, over a dozen people in the Pacific Northwest were killed and others became very ill from eating hamburgers made from meat contaminated with a dangerous strain of *E. coli,* a common bacteria. A century ago in the United States at least 25% of the children died of bacterial infections before the age of 12. Protozoa can cause a variety of diseases, including sleeping sickness, malaria, and amoebic dysentery. Viruses cause many childhood diseases, including flu, mumps, measles, and chicken pox.

# 6

## Detecting Contaminated Water: The Biological Risks

### Introduction

**Investigating Microorganisms**

The microscopic world of living things was unknown just a few hundred years ago. Today it is common knowledge, although most people are aware of only the negative effects of microorganisms. Everyone knows that "germs" cause infection and disease. Parents struggle to teach their young children sanitary habits. But saying that microscopic living things exist is very different from finding direct evidence.

### Challenge

Your challenge is to use a microbial growth medium to prove that a sample of untreated water contains microoganisms.

### Materials

*For each group of four students:*

- Two prepared and sterilized petri dishes
- Two sterile inoculation sticks
- Two small pieces of masking tape for labeling
- One source of untreated water

*Activity 6*

*Detecting Contaminated Water: The Biological Risks*

## Procedure

*Part One: Inoculating the Two Petri Dishes with Untreated Water*

**Safety**

**Never open the petri dishes after they have been inoculated.** Make all observations with the dishes taped closed. Some microorganisms could be harmful.

1. Following your teacher's example, carefully streak two of the growth plates with the untreated water. Minimize or avoid all possible sources of outside contamination.

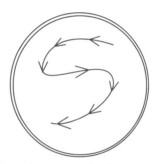

2. Tape the plates closed, and label each one on the bottom. The label should identify the class period, group, time and date of inoculation.

*Part Two: Setting Up a Control*

3. Your teacher will streak another plate with a sterile inoculation stick. This will be the *control*.

*Part Three: Setting Up Your Investigation Report*

4. Observe these growth plates every day for about 4 to 5 days. Make a chart to record observations from each growth plate. Record your observations of each plate on a copy of the chart on page 36 or make one of your own.

*Activity 6*

## Detecting Contaminated Water: The Biological Risks

*Observations of Growth*

Growth Plate Observed _____

Date and Time of Inoculation _____

| Day and Time of Observation | General Sign of Growth | Number of Growth Colonies | Visual Record |
|---|---|---|---|
| | | | ◯ |
| | | | ◯ |
| | | | ◯ |
| | | | ◯ |
| | | | ◯ |

# Detecting Contaminated Water: The Biological Risks

## Introduction

### Biological Risks in the Water Supply

This reading describes some of the health risks of drinking unsafe water and some of the microorganisms that are harmful if present in the drinking water supply.

## Challenge

Your challenge is to describe a situation in which you would be at risk of getting one of these waterborne diseases.

### Typhoid Fever

Typhoid fever is caused by bacteria that enter the body in water or food contaminated with the feces or urine of a carrier of the disease. After 7 to 21 days, the illness begins with fever, sleepiness, headache, and loss of appetite. Increasing weakness and abdominal discomfort develop during the second week, when a rose-colored rash may appear. Intestinal bleeding may occur in the second or third week and can be fatal. Treatment has reduced the death rate from 30% to 2%. In cities, transmission of typhoid fever may occur through food contaminated by polluted water or by food handlers who are carriers. Typhoid fever can be prevented by sanitary disposal of human wastes and by chlorination of drinking water.

### Dysentery

Dysentery is a disease of the intestines that causes painful diarrhea and stools containing blood and mucus. It can be caused by bacteria or small one-celled organisms called *amoebas*. It may be fatal. The symptoms are a fever, abdominal cramps, and diarrhea that last about a week. Treatment consists of drinking large quantities of fluids, but antibiotics are given in severe cases. The disease is transmitted through water or food infected with organisms from the stools (feces) of infected persons and is prevented by good hygiene. Amoebic dysentery is caused most often by drinking water containing the amoebas. It can be treated with several drugs but is hard to completely eliminate and often can reoccur.

### Giardia

Giardia are one-celled organisms that can be found in the small intestine of humans and other animals. Giardia are found in up to 27% of all American streams and rivers and are also common in streams worldwide. A giardia infection does not generally cause death. One kind of the organism, *Giardia lamblia,* can cause diarrhea, cramps, and nausea that can persist if untreated. This condition is highly contagious and is spread through contaminated water or food. Infection with *Giardia lamblia* is often referred to as "backpacker's diarrhea" because drinking inadequately treated water on backpacking trips has led to many cases of the disease.

### Hepatitis

Hepatitis is a disorder involving inflammation of the liver. Symptoms include loss of appetite, dark urine, fatigue, and sometimes fever. The liver may become enlarged and produce substances that give the skin a yellow color. Hepatitis A, once called infectious hepatitis, is the most common cause of acute hepatitis. Usually transmitted by food and water contaminated by human waste, such infections can lead to an epidemic in unsanitary areas. Mild cases of Hepatitis A are treated with bed rest but not with drugs. More serious cases can lead to hospitalization and treatment with steroids. Most healthy people recover completely.

# Detecting Contaminated Water: The Biological Risks

 **Questions**

1. Describe a situation in which you would be at risk of getting one of the waterborne diseases described.

2. Assume you moved to a new country and weren't sure about the quality of the local supply of drinking water. You need to decide whether to drink the local tap water or buy bottled water. How would you decide? List the evidence you would collect and the steps you would use to make your decision.

# Activity 7: Chlorination: How Much Is Just Enough?

## Introduction

### The Milwaukee Story

The risk of disease from biological contamination of drinking water is not just a thing of the past. We now know that Milwaukee's effort to keep its water supply safe failed in 1993, causing many to get sick and the entire city population to be inconvenienced. Read this article, written at the time of the problem.

## Challenge

Read this article and be ready to discuss your ideas about what you think might have gone wrong with Milwaukee's water purification system.

Water treatment plants such as this one serve urban areas.

# Chlorination: How Much Is Just Enough?

### *Milwaukee Illness May Be in the Water*

An intestinal illness has stricken hundreds and perhaps thousands of people in the Milwaukee area during the last few weeks, and the city has now been seized with alarm by health officials' suspicion, first announced late Wednesday, that the cause may be contamination of the municipal water supply.

Although the authorities emphasize that the water system has not been conclusively implicated, they have cautioned Milwaukeeans not to drink tap water, or use it to brush their teeth or wash food, unless they first boil it for five minutes to kill the parasite that is now suspect.

The warning has set off a scramble for bottled water, which was emptied from the shelves of many stores throughout the area today. The Milwaukee public schools, which reported high rates of absenteeism among teachers and children alike as a result of the outbreak, covered their water fountains, offering thirsty students milk or juice instead....

And, like bottled water at grocery stores, intestinal remedies were in short supply at pharmacies. "We Have Imodium A-D," one Walgreen's store here proclaimed on a street sign....

Health officials' suspicion now is that their enemy is cryptosporidiosis, an uncommon ailment caused by a parasite, cryptosporidium, a feces-borne organism that is frequently transmitted in water and inhabits the intestines, most often in animals but sometimes in humans as well.

The officials believe that the parasite may have contaminated the municipal water supply, most likely because of faulty operation by one of the system's two purification plants.

The illness's symptoms...include severe diarrhea, abdominal cramps, fever and vomiting. Otherwise healthy people who are stricken usually recover within 7 to 11 days. But the illness can be life-threatening to people, like AIDS patients, whose immune systems are poor. No deaths from the current outbreak have been reported, however....

*Copyright ©1993 by The New York Times Company.*
*Excerpt reprinted by permission.*

*Chlorination: How Much Is Just Enough?*

### Introduction

## A Century of Chlorination

If you know the dangers of pure chlorine, the idea of drinking it in your tap water may seem crazy at first. But in small enough amounts, chlorine does not make people sick, and most can't even sense it (the concentration is below their sensory thresholds). In the United States, chlorine has been added to public drinking water for almost 100 years in order to reduce the risk of disease from microorganisms. Has it worked?

### Challenge

Read this short history of chlorine in the water supply. Think about what the next 100 years might be like if the practice were suddenly stopped.

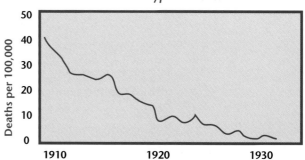

**Deaths in Toronto due to typhoid fever**

# Chlorination: How Much Is Just Enough?

### History of Chlorine in the Water Supply

The chlorination of public drinking water has been called one of the most significant public health practices of the twentieth century. Chlorine was first used in drinking water to remove odors long before its disinfectant powers were known. Because it also acts as a bleach, it removes color from water containing some organic (carbon-containing) chemicals. In the United States, it was first added to the public drinking water supply in 1908 in Jersey City, New Jersey.

As chlorination spread across the U.S., there was a dramatic decrease in the number of deaths caused by cholera and typhoid. Before chlorination was introduced to Toronto in 1910, the yearly death rate from typhoid fever was 44.2 deaths per 100,000 population. By 1928, the death rate had dropped to 0.9 deaths per 100,000 population (see chart).

The incidence of these diseases had almost dropped to zero by the mid-twentieth century. The average life expectancy has increased from 49 years in 1900 to over 75 years today—due in part to an improved understanding of disease and rapid improvements in hygiene, medical practices, and nutrition.

A researcher obtains a water sample for testing.

# 7 Chlorination: How Much Is Just Enough?

## Introduction
### Determining an Effective Level of Chlorination

An effective way to kill microorganisms in water is to add chlorine. But chlorine can be harmful to humans if the concentration is too high. So when treating water with chlorine, it is important to determine the lowest concentration needed to prevent biological contamination. In this lab, you will experiment with green algae, a harmless microorganism.

## Challenge

Determine the lowest concentration of chlorine in ppm that kills green algae.

## Materials

*For each group of four students:*
- One 30-mL dropping bottle of household (5.25%) bleach solution
- One plastic cup filled with about 100 mL of green algae water

*For each pair of students:*
- One SEPUP tray
- One 30-mL dropping bottle of distilled water
- One 50-mL graduated cylinder
- One dropper
- One plastic cup half filled with tap water for rinsing
- One piece of white paper
- One paper towel

Activity 7

# Chlorination: How Much Is Just Enough?

## Procedure

**Safety**

Do not bring chemicals into contact with your eyes or mouth. Wear safety eyewear as directed by your teacher. Avoid getting bleach on your clothing.

1. Set up your investigation report. Prepare a data table like the one on the next page.

2. With your partner, make a four-step serial dilution of the household (5.25%) bleach solution. This solution contains 13,000 ppm chlorine. Begin by putting 10 drops of the chlorine bleach into Cup 1 of the SEPUP tray. Move one drop to Cup 2 and dilute it with 9 drops of the distilled water. Repeat this process until you get to Cup 4.

3. Carefully measure and pour 5 mL of the algae water into each of the five large cups on the SEPUP tray. Be sure to record the color of the algae water when observed over a white background.

4. Treat the water in Cup A with 5 drops of the concentrated chlorine solution from Cup 1. Treat Cup B with 5 drops of the first dilution from Cup 2. Cup C should be treated with 5 drops of the chlorine dilution in Cup 3, and Cup D with 5 drops from Cup 4.

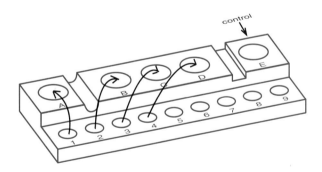

5. The algae water in Cup E is a control. Do not put any chlorine in it. Instead, put in 5 drops of the distilled water.

6. Carefully observe any color changes. Keep track of the time it takes for changes to occur.

7. Complete the Data Processing section.

Activity 7

# Chlorination: How Much Is Just Enough?

*Water Treatment by Chlorination*

| Test Cups (5 mL algae) | Observations (before adding chlorine) | Treatment (5 drops) | Total Chlorine Concentration (ratio 1 to 25) | Observations (after adding chlorine) |
|---|---|---|---|---|
| A | | 25,000 ppm chlorine | 1,000 ppm | |
| B | | | | |
| C | | | | |
| D | | | | |
| E | | | | |

 **Data Processing**

1. Describe what happened in each of the five test cups during the time you observed them.

2. Based on your observations of the effect on algae, what concentration of chlorine would you recommend for water purification? Explain.

3. Is this a fair test for determining the appropriate concentration of chlorine to treat drinking water? Explain.

4. Why do we use algae instead of other microorganisms in our investigations?

5. Why do we add distilled water to the algae in Cup E?

**Activity 7**

*Chlorination: How Much Is Just Enough?*

## Introduction

**Sulfanilamide**

You have just finished examining the risks and benefits of chlorine. Do you ever think about the risks and benefits of prescription drugs? We usually think of prescription drugs as safe if we use them according to the label. It wasn't always that way.

## Challenge

Determine if the Food, Drug and Cosmetics Act (FDCA), which was passed in 1938, is still necessary today.

### Regulating Drugs— The Sulfanilamide Story

Before 1938, American drug companies could make and sell drugs without first trying them out on animals or people, or clearing them with the federal government. It was up to the government to prove that a drug was mislabeled or unsafe before it could be removed from the market. In 1938, the Food and Drug Administration (FDA) was successful in obtaining legislation that required drugs to be cleared for safety before going on the market.

In order to make its case for the need for laws regulating drugs, the FDA presented a variety of stories documenting the ill effects of drugs that were widely available at the time. The case that finally convinced legislators involved deaths produced by a liquid version of the drug sulfanilamide.

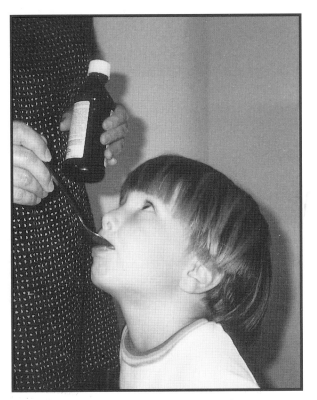

Today, the Food, Drug and Cosmetics Act regulates over-the-counter and prescription medicines.

Activity 7

# Chlorination: How Much Is Just Enough?

This antibiotic was a new wonder drug at the time. It was one of the first antibiotics used to treat bacterial infections. A liquid form of the drug was desired because it was hard for children to swallow in pill form. The company found a solvent for the drug and put the drug on the market without testing. The liquefied version of the drug killed 107 people, mainly children, before supplies were removed from the market.

At the time, the drug was so new that no one knew whether it was the drug or the solvent that was toxic. Further tests showed that it was the solvent, diethylene glycol (used in antifreeze), that was causing the deaths. This tragedy led to the passage, in 1938, of the Food, Drug and Cosmetics Act, which requires products to be tested, usually on animals, and approved for safety before being sold.

### Questions

1. Describe in your Journal what you think would have been an adequate pre-sale test of the safety of liquid sulfanilamide.

2. The Food, Drug and Cosmetics Act (FDCA) is over 50 years old. Is there still a need to keep this legislation on the books? Write a letter to the newspaper expressing your viewpoint in favor of or opposed to the FDCA.

# Section B

## Thresholds, Toxicity, and Risk
## Activities 8 –12

### Overview

You have learned that chlorine can be used to reduce the risk of getting diseases from drinking water. Did you wonder whether there are any risks from the chlorine? In this section of the course, you will learn how investigators measure toxicity and compare risks. This will help you understand evidence and how to make decisions about the risks from chemicals in drinking water and other risks.

To learn about how health risks from chemicals are studied, you will investigate how new medicines are tested before they are sold. New medicines become available all the time. How can we know if these drugs are safe? The study of *toxicity* (*toxic* means poisonous) can provide information about safe levels of drugs and other chemicals that we use. Other studies tell us whether the drugs are effective. For example, one or two aspirin can relieve a headache, but a whole bottle of aspirin can be deadly. A dose of medicine appropriate for an adult might be enough to make a child seriously ill or die. You will build on your understanding of thresholds to understand studies of acute and chronic effects of toxic substances.

Next you will be introduced to the concept of *risk*. All life decisions involve some element of risk. Choosing to take medicine means accepting the small risk that the medicine may have harmful effects. For example, taking aspirin means accepting the small risk that the aspirin will cause stomach problems. Similarly, riding in a car means accepting the risk of an accident. Most of us choose to take medicine and ride in cars because, to us, the benefits clearly outweigh the risks. In the last two activities in this section, you apply the concepts of toxicity and risk. First, in a simulation, you have to choose between two risky decisions. In the final activity, you analyze risk decisions made in relation to a modern day cholera outbreak.

# Activity 8
# Chicken Little, Chicken Big: A Threshold Story

## Introduction

### Chicken Growth Investigation

New substances are constantly being created for possible use in food processing, medicine, and industry. Because many of these chemicals are designed for use with living things, it's important to measure the threshold of effect and the lethal dose.

## Challenge

Use evidence to determine the right amount of growth substance to give a chicken so that it will grow much bigger and stay healthy.

*Chicken Growth Data Chart*

| Age (in weeks) | Average Weight (in grams at end of week) | Food Eaten (in grams during week) | Growth Substance (in grams received during week–range) |
|---|---|---|---|
| Week 0 (hatched) | 40 (when hatched) | | |
| Week 1 | 150 | 120 | 0–6 |
| Week 2 | 400 | 300 | 0–20 |
| Week 3 | 700 | 500 | 0–60 |
| Week 4 | 1,100 | 700 | 0–100 |
| Week 5 | 1,500 | 800 | 0–150 |
| Week 6 | 2,000 | 1,100 | 0–200 |
| Week 7 | 2,500 | 1,300 | 0–250 |
| Week 8 | 3,000 | 1,500 | 0–350 |

*Chicken Little, Chicken Big: A Threshold Story*

## Materials

*For each pair of students:*

- One set of Chicken Puzzle Cards

## Procedure

1. Your team of two will receive a set of 10 cards and a title card. The cards show the results of using different amounts of the growth substance on 10 chickens of the same age for a week.

2. Together you should analyze the data to determine the effect of the growth substance on each of your chickens. There are three possibilities:
   - No observable effect
   - Much heavier growth
   - Death

## Data Processing

1. What relationship can you find between the amount of growth substance used, the beginning weight of the chicken, and the outcome?

2. Based on the results for your group's chickens, what recommendations would you make to the group for the use of this growth substance on chickens? Each of you should record your own recommendations in your Journal.

3. Discuss your recommendations with your group and come to agreement on your group's recommendation to the class.

# 8

## Chicken Little, Chicken Big: A Threshold Story

### Introduction

**The Dose Makes the Poison**

We have seen in the past few activities that using a substance can have both risks and benefits. The amount of the substance is a factor in determining the risk of using the substance. Let's continue our research about the importance of *dose*.

### Challenge

Use your knowledge of threshold from the previous activities to understand the concept of dose.

#### Toxicology: The Dose Makes the Poison

As the celebrated Swiss physician Paracelsus wrote in the late fifteenth century, "All substances are poison; there is none which is not a poison. The right dose differentiates a poison and a remedy."

However, the harmful doses of different chemicals vary widely. Toxicity can be thought of as the potential of a substance to cause illness or even death. Toxicology is the science of poisons— substances that are harmful to living things. The smaller the dose (amount) required to produce a toxic effect, the more poisonous the substance. Even table salt, if taken in sufficient quantities, can be toxic. A ten-year-old child may make a face if she consumes a teaspoonful of salt; 10 teaspoonfuls will make her ill and may even kill her.

Understanding health risks requires an understanding of how the concentration of a substance can affect living things. When the concentration of a substance reaches a certain level (every substance is different), toxic effects begin to occur. The lethal

# Chicken Little, Chicken Big: A Threshold Story

toxicity of a substance is measured in milligrams (one thousandth of a gram) of substance per kilogram of body weight, written mg/kg. Toxicologists have adopted the classification scheme that is shown below for rating the relative toxicity of a substance in humans.

Consider one of the most potent poisons—botulism toxin. Botulism toxin is produced by bacteria in improperly canned food. A single lethal dose is in the range of a hundred thousandth of a milligram per kilogram of body weight. This amount would cover less than 1/100 of the area on the tip of a needle. A thimble full of this toxin could kill the entire population of the earth! Now consider another common chemical—sucrose, or table sugar. There is no evidence of human beings ever receiving a lethal dose. From studies on rats, scientists have determined that the acute lethal dose is 20,000 mg/kg of body weight. You would have to consume more than three pounds of sugar at one time for it to be lethal. This assumes humans respond in the same way as test animals do. This is about as nontoxic as any chemical can be!

*Classification Scheme for Rating Toxicity*

| Toxicity Rating | Probable Lethal Oral Dose for Humans | |
|---|---|---|
| | mg/kg of Body Weight | Volume |
| 1. Practically nontoxic | More than 15,000 | More than 1 quart |
| 2. Slightly toxic | 5,000 to 15,000 | 1 pint to 1 quart |
| 3. Moderately toxic | 500 to 5,000 | 1 ounce to 1 pint |
| 4. Very toxic | 50 to 500 | 1 teaspoon to 1 ounce |
| 5. Extremely toxic | 5 to 50 | 7 drops to 1 teaspoon |
| 6. Super toxic | Less than 5 | Less than 7 drops |

Activity 8

# Chicken Little, Chicken Big: A Threshold Story

### Questions

1. Explain why you agree or disagree with Paracelsus' statement: "All substances are poison; there is none which is not a poison. The right dose differentiates a poison and a remedy." Give some examples to clarify your answer.

2. Review your experience with the powdered drink mix tasting in Activity 2. Did everyone have the same threshold limit for tasting, seeing, or smelling the powdered drink mix? How can that information be used to understand dose?

3. Would the lethal dose of ferrous sulfate (iron supplement tablets) for a 20-kilogram child be the same or different than for a 60-kilogram student? Explain.

4. What are some explanations for why alcoholic drinks affect some people more than others? Give as many reasons as you can.

# Activity 9

## Lethal Toxicity

### Introduction

**Miracle Drug**

A drug that relieves the symptoms of cold, flu, muscle ache, and headache is found to have a *lethal* dose of 300 mg/kg of body weight. The drug is distributed in 325 mg tablets. The *normal dose* prescribed for an average adult (70 kg) is 2 tablets every 4 hours not to exceed 12 tablets a day.

### Challenge

Complete the following questions and then decide if you would take this drug.

### Questions

1. What is the rating for the toxicity level of this drug? (See the table in Activity 8 on page 53 of the Student Book.)
2. What is the lethal dose in mg for a 70-kg adult?
3. What is the normal dose in mg for a 70-kg adult?
4. How many tablets would a normal adult have to take before they reached the lethal dose?
5. Would you take this drug if it were prescribed for you? Explain your answer.

# 9
## Lethal Toxicity

### Introduction
**Acute Toxicity**

You will do a simulated acute toxicity experiment. Your goal is to determine the lethal toxicity level for a single dose of a potentially toxic substance.

### Challenge

Use the simulation to understand how real toxicity experiments are done.

Rats are frequently used in toxicity studies.

Activity 9

# Lethal Toxicity

## Materials

For each group of four students:

- One 30-mL dropping bottle of each of the following:
  - Potentially Toxic Substance (PTS) solution (0.025M HCl)
  - Rat Food (RF) solution (0.1M NaOH)
  - Rat Weight Factor (RWF) solution (Alizarin solution)

For each pair of students:

- One SEPUP tray
- One stirring stick
- One paper towel
- Two pieces of graph paper

## Procedure

**Safety**

Wear eye protection and avoid skin contact with the chemicals.

1. Make all rats the same weight by putting 5 drops of Rat Weight Factor (RWF) solution in each of the cups. Add one drop of Rat Food (RF) to each of the cups.

2. Add drops of the Potentially Toxic Substance (PTS) to Cups 1–8, as shown by the amounts in the figure. Cup 9 has no added PTS; it represents the control group.

*Continued on next page →*

Activity 9

# Lethal Toxicity

**Procedure (continued)**

3. Stir the mixtures after adding all of the solutions. If the resulting mixture for each rat is bluish purple, the rat is said to be healthy. If the mixture is red to orange-yellow, the rat is ill. If the mixture is yellow with no orange in it, the rat is dead.

4. Record the rat health data in a table. Your data table should include the amount of PTS given to each "rat," the color you observe, and your decision about the rat's health. Is it healthy, ill, or dead?

5. Complete the Data Processing section.

## Data Processing

1. What is the minimum number of drops that caused any rats to die?

2. If each drop of PTS solution contains 25 mg of Potentially Toxic Substance, how many mg of PTS are needed to cause death in these rats?

3. Since 5 drops of RWF solution represent a rat weighing 250 grams, calculate the milligrams of PTS per gram of rat to cause death.

4. Calculate how many milligrams of PTS per kilogram of body weight represent the lowest lethal dose (1 kg = 1,000 g).

5. If we assume that human beings react exactly the same as rats to the PTS, and the only difference is due to weight, calculate the milligrams of PTS that would be the lowest lethal dose to a 70-kilogram human (154 lb.).

*Lethal Toxicity*

# 9

## Introduction

### Chronic Toxicity

The following data simulate a chronic toxicity test. There were seven groups, each of which had 60 rats. All of the 420 rats weigh the same and are from the same breed. Group 1 is the experimental control group, which did not receive any of the Potentially Toxic Substance. Groups 2–7 had increasing doses of the toxic material, which was given orally. The substance was given for a two-year period.

## Challenge

What chronic dose of Potentially Toxic Substance causes a significant increase in liver tumors in rats?

### Analyzing the Data

Graph the data to see the relation between the dose given and the number of rats with tumors.

*Chronic Toxicity Test Data*

| Group | Daily Dose mg/kg | Number of Rats | |
|---|---|---|---|
| | | Dead Before 2 Years | Survived with Liver Tumors |
| 1 (control) | 0 | 2 | 4 |
| 2 | 10 | 1 | 4 |
| 3 | 20 | 2 | 3 |
| 4 | 30 | 3 | 4 |
| 5 | 40 | 2 | 10 |
| 6 | 50 | 3 | 15 |
| 7 | 60 | 4 | 19 |

Activity 9

# Lethal Toxicity

 **Data Processing**

1. What is the largest dose that caused no increase in tumors? This is called the no-effect level (NOEL). The threshold value is the area between the no-effect level and the level where negative effects occur. Label the threshold value on your graph.

2. If humans behave similarly to rats, would you expect a daily dose of 1 mg/kg to cause any negative health effects in humans?

3. Would you recommend the use of this substance as an oral asthma medication? Explain your answer.

4. Explain why it is important to study *both* chronic and acute toxicity in the PTS you examined.

5. In the "Chicken Little" activity, you learned that when the percentage of growth substance equaled 5% of chicken's weight, that was the threshold of positive growth effect. The threshold dose for death was found to be 10%. Using this evidence, explain the statement, "The dose makes the poison."

*Lethal Toxicity*

# 9

## Introduction

### Artificial Sweeteners

The average American obtains approximately 21% of his or her total daily calories from natural and added sugar. People between the ages of 12 and 29 drink more regular, sugar-loaded sodas than any other age group. An average 15-year-old boy consumes two per day (equal to 20 cubes of pure sugar) and the average 15-year-old girl consumes one and one-half cans per day (15 cubes of pure sugar). Is all this sugar consumption healthy? Are artificial sweeteners the answer to our sweet needs?

## Challenge

Weigh the benefits and risks of using artificial sweeteners.

The first sugar substitute, saccharin, was discovered in Germany in the late 19th century and became commercially available in 1900. It had no calories, compared to table sugar's 18 calories per teaspoon, and was quickly promoted as a way to lose weight. It also helped diabetics to control their sugar intake. Another artificial sweetener, called cyclamate, was developed in the 1950s but was banned by the federal government in 1970 when it was found that it caused cancer in laboratory rats. Under the Delaney Clause to the Pure Food and Drug Act, no substance that causes cancer in laboratory animals may be used as a food additive for human consumption. But there is one notable exception. In 1977 the United States Food and Drug Administration (FDA) banned the use of saccharin in prepared foods after Canadian scientists found that rats given large amounts developed bladder cancers. Because of the controversy over the test results, Congress reversed the ban but said that all foods containing saccharin must be labeled,

*Lethal Toxicity*

"Use of this product may be hazardous to your health. This product contains saccharin which has been determined to cause cancer in laboratory animals."

With the ban on cyclamates and concerns about the safety of saccharin, a new artificial sweetener called aspartame was approved for use in 1980. It is commonly called Nutrasweet and has replaced saccharin in diet beverages. It is not without its problems, however. Unlike saccharin, which has a bitter aftertaste, aspartame has none, but it can cause severe reactions for people who have a genetic disorder called PKU. You might want to find out more about this and write a short report in your Journal. Another problem with aspartame is that it is destroyed by heat so it cannot be used in most cooked foods.

Some new low-calorie sweeteners are under development and may gain FDA approval. One sweetener already approved in 1988 for limited use in chewing gum and dry milk is Acesulfame K.

*Table of Comparative Sweetness (compared to table sugar)*

| Substance | Sweetness Value |
|---|---|
| Saccharin | 300 |
| Aspartame | 200 |
| Acesulfame K | 130 |
| Cyclamate | 30 |
| **Sucrose (table sugar)** | 1 |
| Fructose | 1.40–1.75 |
| Glucose | 0.70–0.75 |
| Corn Syrup | 0.60 |
| Lactose (milk sugar) | 0.15 |

# Lethal Toxicity

Imagine hearing on the radio the following commercial:

Are you looking for that high-calorie candy bar or piece of chocolate cake as a reward for choking down the vegetables? AND, are you feeling guilty, like many Americans who have a taste for sweets, but no control? Now you can eat all the sweet foods you like with no fear of gaining weight! Just add Alpha Sweet to all your favorite foods. Make desserts fit for royalty. Amaze your friends with their great taste and low, low calories. Pick up a package today at your local supermarket, and find out how sweet it is!

## Questions

1. Are the claims in this ad believable?
2. What kinds of questions would you like to ask the makers of Alpha Sweet before using it?
3. What kinds of alternatives would you like to consider?

# Activity 10: Risk Comparison

## Introduction

### Risks of Dying

How do we decide how much risk is too much? Is zero risk possible? The chart on the next page lists a variety of risks to which people are exposed. For this table the total population of the United States is 240,000,000. As you review the chart, think about how the risks compare.

## Challenge

Select from the chart the two risks you think are the highest and the two you think are the lowest. Record them in your Journals.

A helmet reduces the risks associated with riding on a motorcycle or bicycle.

# Risk Comparison

*Perceived Risks of Death*

| Risk | Population at Risk in U.S. |
|---|---|
| Accidental poisoning | 240,000,000 |
| AIDS | 240,000,000 |
| Air pollution | 240,000,000 |
| Alcohol usage (non-abuse) | Estimated |
| All cancers | 240,000,000 |
| All effects of regular smoking | Estimated |
| Animal and insect bites | 240,000,000 |
| Chlorinated drinking water | Estimated |
| Drowning | 240,000,000 |
| Eating charcoal-broiled steak (3 oz. per day) | Estimated |
| Fires | 240,000,000 |
| Heart disease | 240,000,000 |
| Home accidents | 240,000,000 |
| Homicide | 240,000,000 |
| Hunting | 22,000,000 |
| Lightning | 240,000,000 |
| Living with a regular cigarette smoker | Estimated |
| Motorcycling | 4,156,000 |
| Motor vehicle accidents | 240,000,000 |
| Peanut butter (4 tablespoons per day) | Estimated |
| Playing high school football | 1,000,000 |
| Playing professional football | 1,500 |
| Sport parachuting | 25,000 |
| Swimming | 82,000,000 |

# 10 Risk Comparison

## Introduction

### Age Comparisons

The graphs on the next page present evidence about causes of death for people of different ages. The first graph presents data for three leading causes of death. The second graph focuses on two specific causes of accidental death.

## Challenge

Use the graphs to determine how the risk of dying from each cause is related to age. Try to explain the evidence presented in the graphs.

The number of deaths from motor vehicle accidents varies among different age groups.

Activity 10

## Risk Comparison

*Leading Cause of Death, by Age, 1988*

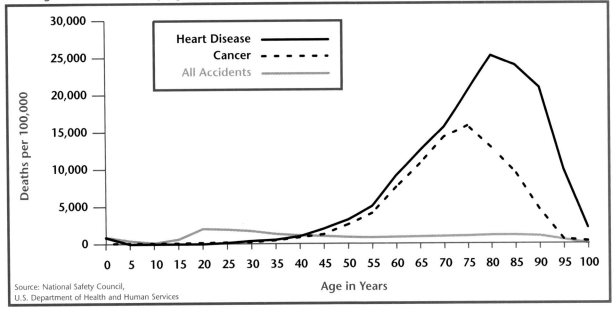

*Causes of Accidental Death, by Age, 1988*

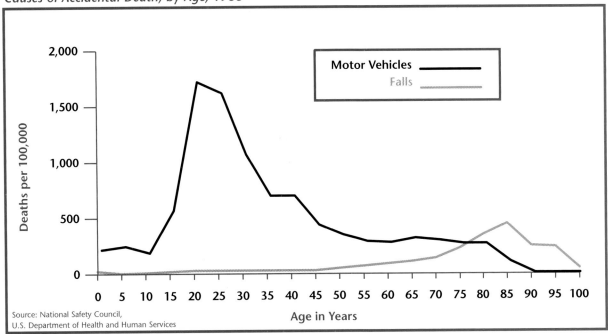

### Question

What conclusions can you draw about the causes of death for different age groups?

Activity 10

# Activity 11

## The Injection Problem

### Introduction

**What is Your Decision?**

In the last activity, you learned about risk analysis. How can risk analysis help us make personal decisions? Read the following information carefully. Imagine yourself in the situation described in the following story.

### Challenge

Use the evidence in the following story to determine whether you would be injected.

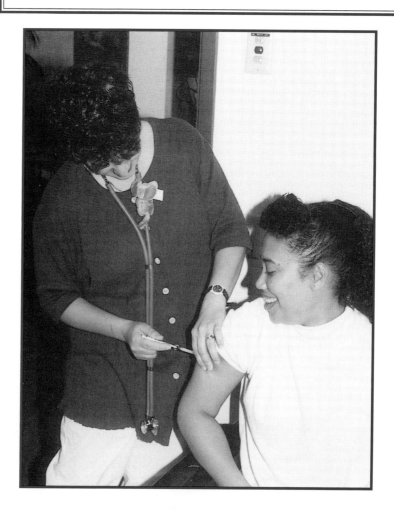

Modern inoculation is required for school attendance.

# The Injection Problem

### Disease Strikes Center City Twice in Five Years

Five years ago a mysterious disease struck Center City, which has a population of 45,000. You live in a small town near Center City. At first, only a few people became sick, running high fevers and breaking out with sores all over their bodies. After a week, many people of all ages became sick in the same way. One month later, the disease disappeared as suddenly as it began. A total of 4,865 men, women, and children caught the disease, and 805 of them died. All of those who survived were left with scars from the sores. Some scars were very bad on people's arms, legs, and faces.

### Some Scientific Detective Work

A year later the disease broke out again in the same city in the same way. When it was over, 5,256 people had caught the disease, and 750 had died. Dr. Martha Stanton, an *epidemiologist* (a person who studies how diseases spread), realized that no one who caught the disease the first time and survived caught the disease the second time. (She could tell because no one who had scars from the first outbreak of the disease caught it the second time.) She reported the information to the medical doctors in the city.

A few doctors and other public health officials proposed the idea that once people survived the disease they were immune and could not get it again. Dr. Walters, a children's doctor, noticed that in homes where children got the disease first, many parents, brothers, or sisters who lived with them caught the disease afterward, but very few died. He studied them further and found out that many of the people who caught the disease this way and survived had cuts or other open wounds when the first family member became sick. Dr. Walters' idea was that people can catch the disease by contact with the material in the infected sores of other victims. His evidence indicated that these people usually did not get a very serious case of the disease.

# The Injection Problem

## A Plan of Action

Dr. Walters reported this to the other doctors in Center City. Five of them decided that if another epidemic of the disease broke out, they would offer to give people the disease. They explained that this meant taking a bit of the material from the sores of someone who had a slight case of the disease and injecting it under the skin of the healthy person. They further explained that almost all people so treated would probably get the disease and that some might even die. They said their idea or hypothesis was that very few people would die and that all who survived the treatment would then be immune.

## Disease Strikes a Third Time

In the next three months, the disease broke out in two other counties; 12,500 people caught it, and 1,594 died. A few months later, people began getting sick again in Center City, and Dr. Walters and five other doctors decided they would give the disease to volunteers in an experiment to test their ideas. In all, 869 people volunteered and were injected. Of these, 863 caught the disease, and 22 died. At the same time, 4,328 other people in Center City caught the disease, and 652 died. Again, no one who previously had the disease caught it again.

There were many arguments in Center City as to whether Dr. Walters and the others who injected people were right to do so. There was another outbreak of the disease in Center City last year, and none of the injected people who survived the earlier outbreak caught the disease. Dr. Walters said this helped prove his idea that injection was a good thing to do.

# The Injection Problem

## Would You Be Injected?

In the last week, 15 people where you live caught the disease. One is in your neighborhood. The Health Department announced yesterday that it expects a serious outbreak of the disease in the area, including your neighborhood. Three doctors in the community are offering to inject anyone in the area who wants to be injected. They explain that injected people will probably get the disease, and some may die. They say it is an experiment and everyone should think about the information from Center City before making up their minds. They think injection will cause fewer deaths, and, as far as they know, those who survive will be immune. You have to make the decision by yourself.

### Questions

1. Would you be injected? What are your reasons for your decision?

2. Assume that all the other evidence about the number of people catching the disease and the number dying remains the same. Would your decision change if 250,000 people lived in Center City? Why or why not?

# Activity 12

## The Peru Story

### Introduction

**Testing for Chlorine in Water**

Chlorine is commonly added to the drinking water supply to prevent growth of bacteria and other microorganisms. As you learned in Activity 7, the effectiveness of chlorine depends on its concentration.

### Challenge

Test the three provided water samples to determine their chlorine concentrations.

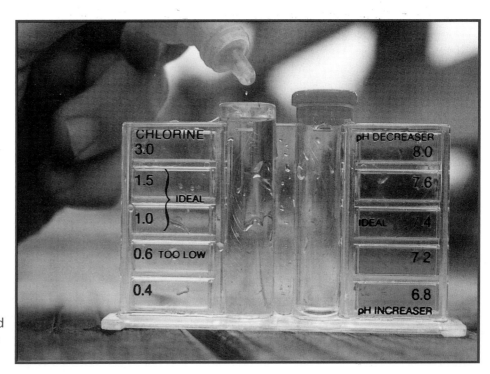

This apparatus is used to test chlorine levels in pool water.

# The Peru Story

## Materials

*For each group of four students:*

- One 30-mL dropping bottle of each of the following:
    - Chlorine test solution
    - Distilled water
    - Household (5.25%) bleach solution
- One chlorine scale comparator
- Samples of pool water and tap water

*For each pair of students:*
- One SEPUP tray
- One dropper
- One stirring stick
- One paper towel
- One piece of white paper

## Procedure

### Part One: Setting Up the Investigation

 **Safety**

Do not taste or smell chemicals or bring them into contact with eyes or mouth. Wear safety eyewear as directed by your teacher.

1. Assemble the SEPUP tray and other materials provided. Place a piece of white paper under the tray.

2. Place 10 drops of bleach solution into Cup 1 of your tray.

3. Perform four step-by-step dilutions of the chlorine solution in Cups 2–5. Make each dilution a 1-in-10 (or 1/10) dilution. Use distilled water for the dilutions.

4. Add 10 drops of pool water to Cup 7.

5. Add 10 drops of tap water to Cup 8.

6. Add 10 drops of distilled water to Cup 9.

*See illustration on next page →*

Activity 12

*The Peru Story*

*Investigation setup*

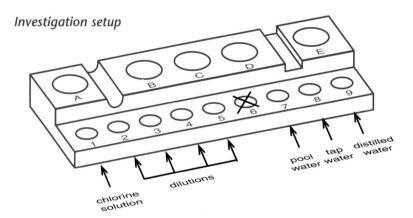

*Part Two: Using the Comparator to Determine the Concentration of Chlorine*

1. Decide how you will record your data.

2. Add one drop of chlorine test solution to Cups 1–5 and 6–9.

3. Hold a chlorine color comparator up to each cup. Compare the color in the cup to the comparator.

4. Estimate the concentration of chlorine in the cup in ppm. Record your estimate.

 **Data Processing**

1. Did you notice any differences between the colors you found in the cups and those on the color comparator? What might cause those differences?

2. Did you compare each chlorine test with a control? If you did, what was the control?

3. The expected values for pool water are between 1.0 and 3.0 ppm chlorine. For tap water they are 0.2 to 0.5 ppm chlorine. How do your results compare with these values?

4. What concerns would you have if the concentration of chlorine in the pool water or tap water was too high? Too low?

# The Peru Story

## Introduction

### Tiny Cancer Risk in Chlorinated Water

You used chlorine to kill microorganisms and know that we use chlorine to purify drinking water. The tests on rats showed us that the dose makes the poison. Is there a danger in using chlorine to purify drinking water?

## Challenge

Examine and react to the possible danger of water chlorination to humans.

## Question

Based on the information in the article, would you recommend chlorinating the drinking water in your community? Give your reasons.

### 700 Extra People May Get Cancer Each Year in the U.S.

Chlorinated drinking water has been linked to small increases in the rates of rectal and bladder cancer in a new analysis by researchers at Harvard University and the Medical College of Wisconsin.

The findings, reported today in *The American Journal of Public Health,* are drawn from a combination of 10 previous studies. Using statistical methods, the researchers found that the slightly higher rates of the two cancers seemed to correlate with the amount of byproducts produced by chlorinated water....

About three-fourths of the water supply in the United States, including that of New York City, is chlorinated. The degree varies widely, depending largely on the source of the water and the contaminants in it. Also, the studies the researchers analyzed were done in the 1970's, when much more chlorine was allowed than now....

Scientists are handicapped in evaluating the benefits and hazards of chlorination because they do not know how many cases of serious infection would occur without it....

...Dr. Morris's team found that the relative risk for bladder cancer was 1.21. For rectal cancer it was 1.38. In other words, people drinking chlorinated water had a 21 percent greater risk of getting bladder cancer and a 38 percent greater risk of getting rectal cancer than those who drank non-chlorinated water....

(Copyright © 1992 by The New York Times Company. Excerpt reprinted by permission.)

# 12

## The Peru Story

### Introduction

**Cholera Epidemic in Peru**

You made a decision about chlorinating the water in your own community. The health officials in Peru in South America who studied the evidence also made a decision. Let's read about their decision and see if this affects your decision.

### Challenge

Read "The Peru Story" and use all the evidence you have collected since the beginning of the course to consider the tradeoffs involved in chlorinating the water supply.

#### The Peru Story

A major outbreak of cholera began in Peru during 1991. Pan American Health Organization officials believe that the bacterium that causes cholera first arrived with a Chinese freighter that dumped its contaminated bilge water into the harbor of Lima, Peru. The organisms quickly spread to fish and shellfish and probably first infected humans in servings of raw fish called *ceviche*. Once humans were infected, the water supply was at risk as individuals put hands into home water storage containers or local wells, introducing the bacterium into the water supply. Cholera bacteria that get into an untreated water supply can thrive and infect many times as many people as might have otherwise been exposed by person-to-person contact.

According to Christopher Anderson's article in the international science magazine *Nature*, one explanation for the spread of cholera in Peru was that health officials in Peru might have misjudged the relative risks of water chlorination

# The Peru Story

on one hand and microbial contamination on the other. Based on United States Environmental Protection Agency studies showing a possible increase in the risk of cancer due to the by-products formed by chlorinated water, officials from Peru halted chlorination of many of Lima's wells. Chlorine is a disinfectant that is capable of killing the bacterium that causes cholera. However, chlorine also can react with the by-products of organic decay present in most water supplies to produce several suspected animal carcinogens, including chloroform.

According to this report, the decision by Peruvian officials to halt chlorination of much of the country's drinking water is being blamed for the outbreak of cholera in January 1991. As of November 1991, more than 300,000 cases of cholera had been reported, and more than 3,500 lives had been claimed by the epidemic. Starting with this outbreak of the disease in Peru, cholera in South and Central America is still a significant problem today.

Another explanation for the spread of cholera suggested by Peruvian officials who disagree with the report in *Nature* was that, although Peru has good water filtration technology and pumps safe water into the drinking water system, old pipes and unchlorinated open wells appear to have allowed the cholera bacteria to enter the water supply after filtration.

(Adapted from Nature, *November 28, 1991, p. 255.*)

### Question

Does the information in this article change your opinion about chlorination of water? Explain.

# 12 The Peru Story

## Introduction
### Chlorination Balance Sheet

The previous articles have given you information about the benefits and risks of chlorination. Assume these articles recently appeared in your local newspaper. The information in the articles has caused some community members to express concern about risks of cancer and waterborne diseases in the drinking water supply.

## Challenge

Use evidence to weigh the tradeoffs involved in water chlorination.

## Assignment

You are a public health official who works in the Water Department. Your supervisor has asked you to respond to the public's concern about water chlorination at the next City Council meeting. Prepare a written response explaining the issues raised in the newspaper articles. Be sure to discuss the advantages and disadvantages of chlorinating drinking water in your response, and then explain your recommendation about whether the water in your town should be chlorinated.

# Section C

## Chemical Testing
## Activities 13–20

### Overview

You have learned that chlorine is deliberately added to the drinking water supply to prevent the growth of organisms that can cause diseases. There are other chemicals that can contaminate drinking water as the result of natural causes or human activity. Based on a desire to ensure safe drinking water, the federal government has set water quality standards. They are:

1. Tastes, smells, and looks good to drink
2. Is well below the threshold for negative health effects from any substances present in the water

Federal water quality standards set upper limits for certain chemicals in the drinking water supply. If the level of any of these chemicals exceeds the limit, then the water is not considered suitable for drinking.

Setting drinking water standards involves using information about toxicity and risk as discussed in the previous section. There are several methods used to determine the health risks of chemicals. They are often determined by conducting animal tests, similar to the ones you studied in Activities 8 and 9. Sometimes the health risks of chemicals can be studied by looking at their effects on people who have been accidentally exposed to them. Based on the results of these investigations, limits are set for drinking water that should result in very low risk.

To decide whether water meets drinking water standards, we need to know not only *what* chemicals are present in the water, but also *how much* of each chemical is present. Remember, the dose makes the poison. Low levels of minerals are what make spring water taste so much better than distilled water. Higher levels of some of the same minerals might cause health problems.

In this section of the course, you will learn some of the basics of chemical testing. This will help prepare you to answer the question, "Is this water safe to drink?"

# Activity 13
# Chemical Detective

## Introduction

### Chemical Detective

You can identify chemicals by observing their properties and by studying their interactions with other chemicals. For example, gold can be distinguished from other metals by its color and the fact that it does not interact with most acids or other chemicals. The use of chemical properties and interactions to identify chemicals is called *qualitative analysis*.

## Challenge

Your challenge is to identify unknown chemical solutions. First you will study the behavior of four known solutions when mixed with two test solutions called *reagents*. Then, using the data you have collected, you will identify unknown solutions by comparing their behavior to the behavior of the four known solutions.

Students test for heavy metals.

# Chemical Detective

## Materials

*For each group of four students:*
- One 30-mL dropping bottle of each of the following:
  - Solutions 1, 2, 3, 4
  - Reagents A and B

*For each pair of students:*
- One plastic bag with "Chemical Detective Grid" inside
- One paper towel

## Procedure

### Safety

Use caution when working with chemical solutions. Wear safety eyewear and avoid spilling the solutions on your skin. Reagent A, the silver nitrate, will stain your skin and clothing. Be sure to wash your hands at the end of the activity.

1. Set up your investigation report form. You will need a data table for recording your results. Your team will be testing seven different solutions with two different testing solutions (Reagents A and B).

2. On the plastic-covered grid, combine one drop of a numbered solution (Solutions 1–4) with a lettered reagent solution in the corresponding space on the grid. *Be careful not to touch the tip of the dropping bottles to the drops of solution.*

   (For example, in the first row of the grid shown below, both spaces should get a drop of Solution 1. A drop of Reagent A should be added to the drop of Solution 1 in the first space in the row. A drop of Reagent B should be added to the drop of Solution 1 in the second space in the row.)

*Qualitative Analysis*

| Numbered Solutions | Lettered Solutions (Reagents) | |
|---|---|---|
| | A (silver nitrate) | B (phenolphthalein) |
| 1. sodium hydroxide (NaOH) | | |

*Continued on next page →*

Activity 13

## Chemical Detective

**Procedure (continued)**

3. Record your observations on your investigation report. *Do not clean up the drops until you have finished testing your unknowns.*

4. Have your partner place a drop of one of Solutions 1–4 in each space in Row 5 (First Single Unknown) on your plastic grid without letting you see which solution it is. **This is your unknown.** Use Reagents A and B to test your unknown, and record your results in your data table. Then switch places with your partner. *Place a drop of one of the numbered solutions in the spaces in Row 6 (Second Single Unknown) for your partner to test an unknown solution.*

5. Answer Question 1 of Data Processing in your SEPUP Journal.

6. Your teacher will place on Row 7 of your grid two drops of a combination unknown solution. It contains a mixture of two solutions selected from Solutions 1–4. Record the unknown number (I, II, or III) on your data table. Test the combination unknown solution and record your results in your data table.

7. Answer Questions 2–4 in your SEPUP Journal.

8. Clean the plastic-covered grid by wiping it with a damp paper towel. Dry the plastic cover.

# Chemical Detective

 **Data Processing**

1. What do you think your unknown solution is? How do you know? Be specific.

2. On the basis of the evidence obtained using Reagent A, which (if any) of the four solutions cannot be present in the combination unknown solution? Which of the four solutions could be present?

3. On the basis of the evidence obtained using Reagent B, which of the four solutions cannot be present in the combination unknown? Which could be present?

4. What do you think your combination unknown contains? Explain how you used the evidence you described for Questions 2 and 3 to reach your conclusion. Be specific.

# Activity 14

## Acids, Bases, and Indicators

### Introduction
**Acid Waste: An Environmental Issue**

In the last activity, you were introduced to *qualitative analysis*, which is used to find out what kinds of chemicals are present. There are many kinds of chemicals that are of concern in determining water quality. In the next few activities, you will learn about one important class of chemicals—acids and bases.

### Challenge

Read and analyze the following newspaper article. As you read it, think about what evidence and information you need to understand the issue described in the article.

Students investigate acids and bases.

*Acids, Bases, and Indicators*

# 14

## SEPUP Sentinel

### City Officials Meet with C-Chip

Recent protests by citizens may hold up the construction of a computer chip manufacturing plant in Endicott City. The C-Chip Company has obtained a permit to build the plant in West Endicott near the Endicott River.

Protesters state that acids produced by the plant will be dumped in the river and harm wildlife in the region. According to Jim Gallagher, representative of the West Endicott Coalition for the Urban Environment, the wastes released by the plant may "throw off the natural pH balance or the amount of salt in the river downstream from the plant. This could damage ecosystems in the river and the marsh where the river feeds into Great Lake."

The computer chip manufacturing process requires the use of strong acid to produce the tiny electrical circuits on the chips. Vera Jackson, Environmental Consultant for C-Chip, stated, "The acid waste will be diluted and neutralized before it is released to the river. This neutral waste will have little or no impact on the diversity of wildlife in the river."

City officials are meeting with officials from C-Chip to discuss the impact of the proposed plant on the Endicott River.

### Questions

1. Write down any words from the article that you don't understand.

2. List questions you would like answered or information you would like to have before making a decision about the issue described in the article.

*Activity 14*

# 14 Acids, Bases, and Indicators

*Federal Water Quality Standards (1993)*
*Primary Standards for Potable (Drinking) Water*

|  | parts per million (ppm) | parts per billion (ppb) |
|---|---|---|
| **1. Aesthetic Standards** | | |
| Chloride | 250.0 | 250,000 |
| Copper | 1.0 | 1,000 |
| Iron | 0.3 | 300 |
| Total Dissolved Solids | 500.0 | 500,000 |
| Calcium-Magnesium | No Standard | |
| Hardness | No Standard | |
| Sodium | No Standard | |
| | | |
| **2. Primary Health Standards** | | |
| Fluoride | 4.0 | 4,000 |
| Lead | 0.005 | 5 |
| Nitrate | 10.0 | 10,000 |
| Mercury | 0.002 | 2 |
| Radon | No Standard | |
| Benzene | 0.005 | 5 |
| Chlordane | 0.002 | 2 |
| Glyphosphate | 0.7 | 700 |
| Total Trihalomethanes (THMs) | 0.10 | 100 |
| Trichloroethylene (TCE) | 0.005 | 5 |
| | | |
| **3. Other Health-Related Standards** | | |
| pH (units) | 6.5–8.5 | |
| Turbidity (NTU) (n) | 1.0 | |
| Coliform Bacteria (% positive sample) | 5.0 | |

# Acids, Bases, and Indicators

## Introduction | Developing a Definition for Acids and Bases

In "Chemical Detective" you learned that some substances can be used to test for other substances. These substances are called *indicators* because they indicate (or show) that another kind of substance is present. One of the substances you used, Reagent B, is an indicator called phenolphthalein. In this activity you will learn how to use phenolphthalein and other indicators to identify a group of substances called *acids* and *bases*.

## Challenge

You will use acid-base indicators to investigate a variety of liquids. You will use your observations to develop a definition that you can use to identify acids and bases.

## Materials

For each group of four students:
- One pH scale or color chart
- Samples of household substances
- One 30-mL dropping bottle of each of the following:
    - Solution 1—1% sodium hydroxide (NaOH) solution
    - Solution 4—1% acetic acid ($CH_3COOH$) solution
    - Universal Indicator solution
    - Reagent B—0.1% phenolphthalein solution

For each pair of students:
- One SEPUP tray
- One stirring stick
- One dropper
- Nine pieces (half strips) of pH paper
- One paper towel
- One 30-mL dropping bottle filled with water

Activity 14

# Acids, Bases, and Indicators

## Procedure

**Safety**

Wear eye protection and avoid skin contact with the chemicals.

1. Set up your investigation report. You will test nine liquids with three different indicators. Think about how to set up a data table to record the information from these tests. Your teacher may have some suggestions.

2. Using the diagram below, set up your tray as follows:

    a. Add 5 drops of water to Cups 1–3.

    b. Add 5 drops of Solution 4 to Cups 4–6.

    c. Add 5 drops of Solution 1 to Cups 7–9.

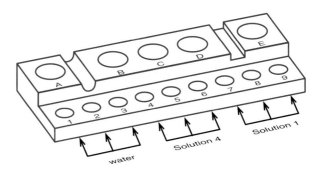

3. Test each liquid with three indicators as follows:

    a. Use one drop of phenolphthalein (PT) to test the liquids in Cups 1, 4, and 7.

    b. Use one drop of Universal Indicator (UI) to test the liquids in Cups 2, 5, and 8.

    c. Test the liquids in Cups 3, 6, and 9 with pH paper.

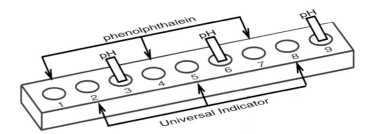

Activity 14

# Acids, Bases, and Indicators

4. Using your data table, record the colors you observe. Rinse and dry your trays.

5. Your teacher will provide various household liquids for testing. Choose any three liquids for your first set of tests. Follow the same testing procedure that you used in steps 2 and 3 to test each liquid with the three indicators. Record the colors you observe. Rinse and dry the tray.

6. Follow the same procedure to test another three household liquids. Record the colors you observe before cleaning the tray.

## Data Processing

1. Use the results you observed to put the nine substances you tested into groups based on how they interact with the indicators.

2. Which seems to be the most useful indicator? Explain your answer.

# Activity 15
# Serial Dilutions of Acids and Bases

## Introduction

### Investigating Acid and Base Dilutions

You have observed that:

- There are indicators that can tell you whether a substance is an acid or base.
- Some indicators are more useful over a wide range while others can only be used over a more narrow range.

For example, phenolphthalein can only be used to detect bases; it cannot distinguish between neutral substances and acids. In contrast, Universal Indicator and pH paper are wide-range indicators, providing information throughout the range from extreme acid to extreme base. You will now examine how you can use a standard scale to estimate the pH number of a solution.

## Challenge

Set up a serial dilution to see how indicators behave as the concentration of acid and base changes. Look for patterns in the behavior of serial dilutions of acid and base when they interact with Universal Indicator.

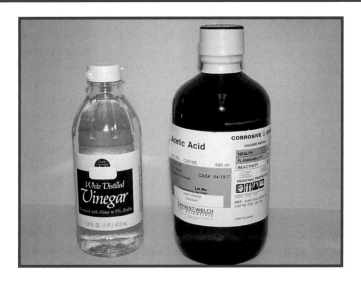

The main active ingredient in vinegar is acetic acid.

# Serial Dilutions of Acids and Bases

## Materials

*For each group of four students:*

- One 30-mL dropping bottle of each of the following:
  - Distilled water
  - Universal Indicator solution
- One pH color chart
- One 30-mL dropping bottle of either:
  - 1% hydrochloric acid (HCl) or
  - Solution 1—1% sodium hydroxide (NaOH) solution

*For each pair of students:*

- One SEPUP tray
- One stirring stick
- One dropper
- Seven pieces (half strips) of pH paper
- One piece of white paper
- One paper towel

# 15 Serial Dilutions of Acids and Bases

## Procedure

**Safety**

Do not taste or touch the solutions. Use safety eyewear. Wash your hands after completing the activity.

1. Set up your investigation report and data table similar to the example shown below.
2. Put 10 drops of water in Cup 7.
3. Perform a serial 1/10th dilution of the solution you are testing in Cups 1–6. If necessary, refer to Activity 3 for the dilution procedure. Put water in one of the large cups for rinsing the dropper and stirring stick between dilutions.
4. Test the pH of each solution with pH paper. Record the results.
5. Test the pH of each solution with Universal Indicator. Record the results.

*Dilution of Solution*

| Cup | Dilution | Color and pH | |
|---|---|---|---|
| | | pH Paper | Universal Indicator |
| 1 | 1/100 | | |
| 2 | 1/1000 | | |
| 3 | | | |
| 4 | | | |
| 5 | | | |
| 6 | | | |

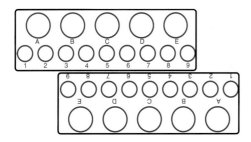

6. Arrange the tray of acid and the tray of base so that the small cups are arranged from most acid to most base, with the neutral cups in the middle. To do this, you will have to place the trays end to end. You may wish to have them overlap where they are neutral.
7. Complete the Data Processing section.

# Serial Dilutions of Acids and Bases

 **Data Processing**

Refer to the results of your group's investigation and the information in the following reading on "Acids, Bases, and pH" to answer the following questions.

1. Draw a diagram of your trays that you arranged from most acid to most base.

    a. Label the cup that contains the known standard with pH 7.

    b. Draw an arrow that shows the direction of neutral pH to strong acid.

    c. Draw an arrow that shows the direction of neutral pH to strong base.

    d. Label each cup with the color and approximate pH of the solution.

2. Write a brief description that you could use to teach someone how the color changes you observed with your serial dilution of acid or base relate to pH change.

# 15 Serial Dilutions of Acids and Bases

## Introduction

### Acids, Bases, and pH

The pH of a solution is a value expressing the solution's acidic or basic nature. A pH value may fall anywhere on a scale from zero (strongly acidic) to 14 (strongly basic), with a value of 7 representing neutrality.

## Challenge

Relate the information in the reading on the next page to your observations of pH in your serial dilutions of acid and base.

*pH Survival Range for Aquatic Organisms*

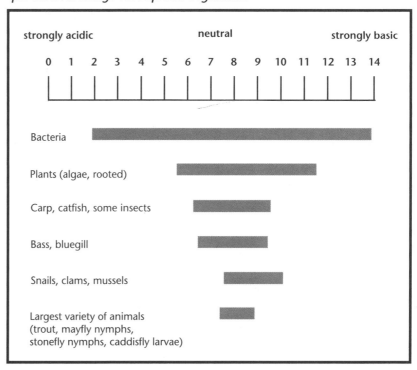

Activity 15

# Serial Dilutions of Acids and Bases

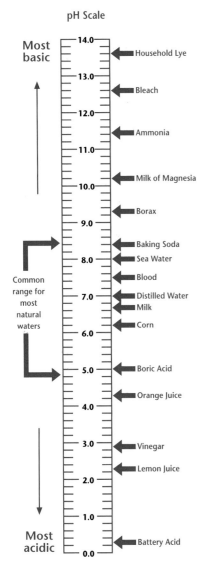

pH Values of Some Common Substances

The measurement and control of pH is important in the manufacture of foods, paper products, and chemicals. In agriculture, testing and maintenance of soil pH is necessary for good yields of crops such as wheat, barley, corn, and other fruits and vegetables. Measurement of pH is also used in studying acid rain and maintaining water quality.

The federal government's standard for drinking water requires the water to have a pH from 6.5 to 8.5. If drinking water has too much acid or base, it may be harmful to human health or it may react chemically with the water pipes (which are usually made of metal) to result in contaminated drinking water. In the environment, many aquatic animals and plants cannot survive in pH levels below 6 or above 8.5.

Using indicators—substances that change color with a change in pH—is a way to measure pH. The color changes are the result of chemical reactions between the indicator and the acid or base.

Acids and bases, and the salts produced by the reaction between acids and bases, are major groups of chemical substances. Originally, acids were recognized by their sour taste in water and because they could attack and dissolve some metals. They also turn pH paper red. Bases turn pH paper blue or violet. Base solutions are usually slippery to the touch and react with acids in water to form salts.

Acids and bases are important in the manufacture of many products. Sulfuric acid is by far the largest single product of the chemical industry. It is used in manufacturing fertilizer, refining petroleum, and pickling metals. Nitric acid is used in manufacturing explosives and dyes. Of the bases, sodium hydroxide and potassium hydroxide are used in soapmaking, and ammonia is used as a fertilizer and in the production of other fertilizers.

## Questions

1. Why is the pH of water important?
2. Explain how the dilutions in your trays can be used as a model for the pH scale shown on this page.

# ACTIVITY 16

# Acid-Base Neutralization

## Introduction

**Mixing Acids and Bases**

Previous investigations have produced evidence that

- Indicators can be used to determine whether solutions are acid, base, or neutral.
- There is a range of indicator colors, depending on the concentration of the acid or base solution.
- As acids and bases are diluted with water, they interact with the indicator more like water.

Now you will begin to investigate what happens at the molecular level when solutions become neutral.

## Challenge

Use your observations to help you develop a model for neutralization.

## Materials

*For each group of four students:*

- One 30-mL dropping bottle of each of the following:
    - Solution 4–1% acetic acid ($CH_3COOH$) solution
    - Solution 1–1% sodium hydroxide (NaOH) solution
    - Distilled water
    - Universal Indicator solution

*For each pair of students:*

- One SEPUP tray
- One stirring stick
- One piece of white paper

# Acid-Base Neutralization

## Procedure

**Safety**

Wear safety eyewear. Avoid spilling the solutions on your skin. Be sure to wash your hands at the end of the activity.

1. Set up your investigation report.
2. Place 10 drops of Solution 1 (base) in Cup 1 and 10 drops of water in Cup 9.
3. Add one drop of Universal Indicator to Cups 1 and 9.
4. Add Solution 4 (acid), one drop at a time, to the base in Cup 1. Stir after each drop. Record the colors you observe after each drop is added. *Be sure to record the number of drops that turn the base the same color as the water in Cup 9.* Stop when the mixture turns red.
5. Record on the class bar chart the number of drops of acid you needed to neutralize the base.
6. Place 10 drops of Solution 4 (acid) in Cup 5.
7. Add one drop of Universal Indicator to Cup 5.
8. Add Solution 1 (base), one drop at a time, to the acid in Cup 5. Stir after each drop. Record the colors you observe after each drop is added. *Be sure to record the number of drops that turn the acid the same color as the water in Cup 9.* Stop when the mixture turns blue.
9. Record on the class bar chart the number of drops of Solution 1 (the base) needed to neutralize the acid.
10. Follow your teacher's instructions for reporting your findings.

## Data Processing

1. Which solution seems stronger, the acid or the base?
2. Explain your answer to Question 1.

Activity 16

# 16

## Acid-Base Neutralization

### Introduction — A Model for Acid-Base Neutralization

You have observed that when you mix an acid and a base in just the right amounts, the resulting solution has neither acidic nor basic properties when tested with an indicator. The resulting solution is *neutral*. How can we explain this observation?

Sometimes it is helpful to make a model to understand how something works. When an architect draws plans for a building, he or she may make a small model of it to show people what it will be like. The architect's drawings and blueprints are also models of the building. A model can be a physical model—like a miniature copy of a building, a diagram—like an architect's blueprints, or even a mathematical equation.

When you acted out the role of a drop of acid or base solution, you were creating a model to help explain acid-base neutralization. You represented the drop of solution, and blue or red dots represented particles (molecules) of acid or base. You modeled neutralization when you had equal numbers of acid and base particles come together.

### Challenge

You will make another kind of model for the acid-base neutralization. You will use your models to help you answer some questions about acid and base particles.

Activity 16

# Acid-Base Neutralization

## The Model

In the following diagrams, let each shape of a drop represent one drop of solution and each triangle represent a particle of acid or base. Suppose, as shown here, each drop of acid contains the same number of particles as each drop of base. Then we can see that when one drop of base is added to one drop of acid, we have combined equal numbers of particles of acid and base.

Acid          Base

If, however, the acid is twice as concentrated as the base, then each drop of acid has twice as many particles as each drop of base. Therefore, 2 drops of base are needed to neutralize one drop of acid.

Acid          Base

# 16 Acid-Base Neutralization

**Questions** *(answer in your Journal)*

1. According to the bar chart from the class data, are there more particles of acid in a drop of Solution 4 (acid) or more particles of base in a drop of Solution 1 (base)?

2. Explain your answer to Question 1.

3. In your Journal, illustrate drops of solution in which each drop of base contains twice as many particles as each drop of acid. Use △ for acid particles and ▲ for base particles.

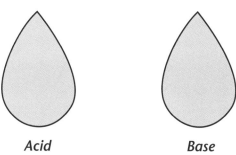

Acid        Base

4. According to the diagram in Question 3, how many drops of acid would be needed to neutralize 10 drops of base? Explain.

5. In your Journal, illustrate drops of solution in which each drop of acid contains three times as many particles as each drop of base. Use △ for acid particles and ▲ for base particles.

Acid        Base

Activity 16

## Acid-Base Neutralization

6. In the example in Question 5, how many drops of base would be needed to neutralize 25 drops of acid?

7. In the example in Questions 5 and 6, how many liters of base would be needed to neutralize

    a. one liter of acid?

    b. 10 liters of acid?

    c. 20 liters of acid?

8. A student conducts an investigation to see which is more concentrated, a sample of acid or a sample of base. He finds that it takes 4 drops of acid to neutralize one drop of base. Which is more concentrated, the acid or the base? Explain your answer. You may wish to include a diagram.

# ACTIVITY 17

## Quantitative Analysis of Acid

### Introduction

You have observed how Universal Indicator and pH paper can be used to give a rough estimate of the concentration of an acid or base. For example, different colors are obtained with these indicators and 1%, 0.1%, and 0.01% acid or base. However, pH indicators cannot distinguish between 1% and 2% acid or base. When we need a more exact measure of the level of an acid or base in a solution, we need to use a different kind of test. This kind of test, which tells us about the amount, or *quantity*, of an acid or base, is called a *quantitative test*.

### Challenge

Your challenge is to quantitatively determine the concentration of an unknown acid sample.

### Materials

*For each group of four students:*

- One 30-mL dropping bottle of each of the following:
    - 5% acetic acid ($CH_3COOH$) solution
    - Solution 1—1% sodium hydroxide (NaOH) solution
    - Distilled water
    - Reagent B—0.1% phenolphthalein solution

*For each pair of students:*

- One SEPUP tray
- One stirring stick
- One piece of white paper

*For each student:*

- Graph paper

# Quantitative Analysis of Acid

## Procedure

**Safety**

Do not taste or touch the solutions. Follow your school policy regarding the use of safety eyewear. Wash your hands after completing the activity.

1. Set up your investigation report form.
2. Add 2 drops of 5% acetic acid ($CH_3COOH$) to small Cups 1, 2, and 3. Then add 8 drops of water to each of these cups.
3. Add 6 drops of 5% acetic acid ($CH_3COOH$) to small Cups 4, 5, and 6. Then add 4 drops of water to each of these cups.
4. Add 10 drops of 5% acetic acid ($CH_3COOH$) to small Cups 7, 8, and 9.
5. Calculate the final % of acid in each cup. Record the percentages you calculate on your investigation report.

    For example, in Cup 1, 2 drops of 5% acetic acid ($CH_3COOH$) were diluted with 8 drops of water. This means that 2 drops out of 10 drops, or 2/10, of 5% acetic acid was used:
    $$(2/10) \times (5\%) = 1\%$$

6. For Cups 1–9, add one drop of Reagent B (0.1% phenolphthalein solution) to each cup.
7. One drop at a time, add Solution 1 (base) to Cup 1. Stir after each addition. Continue until the solution remains pink after stirring. It may be very pale pink or deep pink. Record the number of drops added.
8. Continue the procedure in Step 7 for each of the 9 cups. Be sure to record your results.
9. Calculate the average number of drops needed to neutralize the acid in Cups 1–3, Cups 4–6, and Cups 7–9. Record your calculations. Prepare a data table to summarize your results.
10. Prepare a *line graph* of the average number of drops needed to neutralize each concentration of acid. This graph is called a *standard graph* because it shows the results obtained with standard (known) concentrations of acid.
11. Do not clean up until you have completed Step 2 of the Data Processing section.

# Quantitative Analysis of Acid

 **Data Processing**

1. Explain how you could use the standard graph and procedure you just followed to find the concentration of an unknown sample of acetic acid. (Hint: How many drops of unknown would you need? What would you add to the unknown? How would this help you find out the unknown's concentration?)

2. Obtain some unknown acetic acid (Unknown X or Unknown Y) from your teacher. Carry out the procedure you described in Question 1 above. Use one of the large cups. Be sure to record your result. Explain how it tells you the concentration of the unknown acid.

3. Use your standard graph to predict how many drops of base would be needed in a similar experiment to neutralize 10 drops of 2.5% acetic acid or 10 drops of 4.5% acetic acid. Record your predictions. This procedure is called *interpolation* of data because you are determining the concentrations of solutions that fall *between* known points. (*Inter-* is a prefix that means between.)

4. Use your standard graph to predict how many drops of base would be needed in a similar experiment to neutralize 10 drops of 6% acetic acid. Record your prediction. This procedure is called *extrapolation* of data because you are determining the concentrations of solutions that fall *outside* the known points. (*Extra-* is a prefix that means outside.)

5. Which procedure (extrapolation or interpolation) would you have more confidence in? Explain your answer.

# Activity 18

## The "Used" Water Problem

### Introduction

Imagine you work as the Safety Officer for the C-Chip plant that you read about in Activity 14. The plant uses acid solutions in a process that coats pieces of metal with chromium or gold. After it is coated, acid is washed off the metal. This "used" water has an unknown amount of acid in it.

The plant was built on the banks of a large river so that fresh water would be readily available and "used" water could be dumped back into the river. The law says the "used" water cannot be put back into the river unless it is neutralized. The pH level must fall within the range of natural waters. You are responsible for disposing of the "used" water without causing an environmental problem.

### Challenge

Your job is to neutralize the pH level of the "used" water so that it matches the pH level of the river water. You and your partner will be given small amounts of the "used" water and the river water. Using these small amounts of water, you must determine how much base will be needed to neutralize 100 liters of the "used" water. Keep careful notes of your procedure and your results. You will use them to write up a full report of your investigation.

# 18

## The "Used" Water Problem

### Materials

For each group of four students:

- One 30-mL dropping bottle of each of the following:
  - "Used" water
  - Water (Use this to represent river water before it is used.)
  - Solution 1—1% sodium hydroxide (NaOH) solution
- Your choice of indicator (Universal Indicator or pH paper)

### Procedure

**Safety**

Do not taste or touch the solutions. Follow your school policy regarding the use of safety eyewear. Wash your hands after completing the activity.

**I. Planning Your Investigation**

- Set up a page for your investigation report in your Journal.

- Think about how you might accomplish your task. Make notes in your investigation report.

- Share ideas with your partner. Working together, come up with a plan for finding out how much base is needed to neutralize a small sample of the acid waste water. Be sure to decide how many drops of "used" water you will test. Record your plan. Be sure to list all the materials you will need.

**II. Carrying Out Your Investigation**

- Perform your investigation. If you make any changes to your original plan, be sure to record them.

- Remember to record all results.

# The "Used" Water Problem

### III. Managing Your Data

- Based on your results, calculate how much base you would need to neutralize 100 liters of used water. Record your calculations.

### IV. Preparing a Written Report

- As Safety Officer, you must write a report on your investigation. Start your report on a clean sheet of paper. Include your name and your partner's name, the date, and a title for your report. Your report should include the following four components:

    - A statement of the problem you were trying to solve. In this case, include an explanation of why it is important to be sure to add the correct amount of base to the acid waste water.

    - A description of the materials and procedure you used to solve the problem.

    - A clear presentation of the results you obtained.

    - An analysis of the results and your recommendation for neutralizing the waste. Be sure to include any calculations you performed. You should discuss any problems you had with the experiment, or any sources of error you may have had. Be sure to discuss water quality problems that could occur if you made an error in your experiment or your calculations.

# Activity 19: Is Neutralization the Solution to Pollution?

## Introduction

In Activity 18 you discovered that a base can be used to neutralize acidic waste water. A pH test will tell you that the neutralized "acidified waste" water behaves like distilled water. Does this mean the neutralized waste contains only water? It is important to know if there is a product of neutralization other than water.

## Challenge

Your challenge is to investigate the properties of the neutralized solution to obtain more evidence about the product of the interaction of the acid and base solutions.

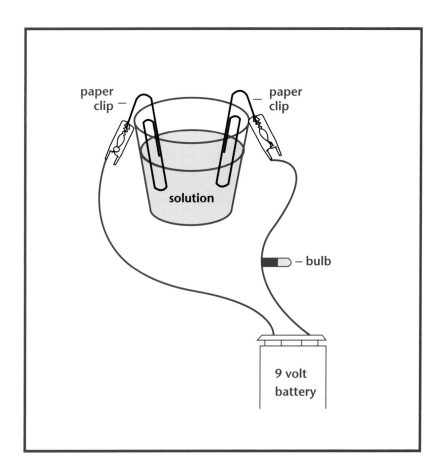

Apparatus for testing conductivity.

# Is Neutralization the Solution to Pollution?

## Materials

*For each group of four students:*

- Two SEPUP trays
- Two stirring sticks
- Two 30-mL graduated cups
- Two paper clips
- One small bulb with attached battery harness
- One 9-V battery
- Water
- One sugar packet
- One salt packet
- 1% hydrochloric acid (HCl) solution
- Solution 1—1% sodium hydroxide (NaOH) solution
- pH paper
- Two droppers
- One 8-ounce plastic cup half filled with water

# 19

## Is Neutralization the Solution to Pollution?

### Procedure

**Safety**

Do not taste or touch the solutions. Follow your school policy regarding the use of safety eyewear. Wash your hands after completing the activity.

1. Set up your investigation report.

2. Attach the battery harness and bulb to the 9-volt battery. Touch the two clips at the ends of each lead together; notice how brightly the bulb shines. **Note:** To avoid wearing out the battery, do not allow the clips to continue to touch each other except when performing the experiment.

3. Set up your SEPUP trays according to the following chart. Each pair of students should be responsible for one tray setup and testing. Each student will record all results.

|  | Tray 1 |  | Tray 2 |
|---|---|---|---|
| Cup A | 10 mL hydrochloric acid | Cup A | 10 mL water |
| Cup B | 10 mL Solution 1 | Cup B | 10 mL sugar-water* |
| Cup C | 5 mL hydrochloric acid 5 mL Solution 1 | Cup C | 10 mL salt-water** |

\* Add a small packet of sugar to 10 mL of water and stir. Use **all** the sugar.
\*\*Add a small packet of salt to 10 mL of water and stir. Use **all** the salt.

4. Use the pH paper and dropper method to find the pH of each solution. Make sure the solution in Cup C is neutral. If it is not neutral, add the appropriate number of drops of hydrochloric acid or sodium hydroxide (Solution 1) until it is neutral. Record the pH of each solution.

5. Attach a paper clip to each lead as shown on page 108. These paper clips allow you to test the solutions without touching them directly to the harness clips. Test each solution for conductivity by placing the two paper clips in the solution and observing the light bulb. Do not allow the paper clips to touch each other when you are testing a solution. Be sure to rinse the paper clips with water in the 8-ounce plastic cup between each test. Record all results. (A sample data table is on the next page.)

# Is Neutralization the Solution to Pollution?

### pH and Conductivity of Liquids

| Solution | pH | Conductive? |
|---|---|---|
| acid | | |
| base | | |
| neutralized acid + base | | |
| water | | |
| sugar-water | | |
| salt-water | | |

## Data Processing

1. Which substances were neutral?
2. Which substances were conductive?
3. Did the neutral acid and base mixture behave like water? Explain.
4. Based on your results, what do you think the product of neutralizing an acid with a base could be? Explain your evidence for your idea.
5. How could you test your idea?
6. Carefully observe the demonstration your teacher will do. Use the information you obtain from the demonstration to support or revise your hypothesis about the neutral solution.
7. Based on everything you have learned about acid-base neutralization, do you think neutralization can be used as a solution to acid-base pollution? Explain your answer thoroughly. Describe both the advantages and disadvantages and other factors that must be considered before reaching a conclusion.

# Activity 20
# Water Quality

## Introduction
### Water Quality Testing

The Silver Oaks Water Department routinely tests water from three important water sources in the community. These three sources are:

- *River water* from the Fenton River that runs along the north side of town. This water is used for swimming, boating, and irrigation of farmland, but not for drinking.

- *Well water* obtained from a test well near the center of Silver Oaks. This well is used to monitor underground water that supplies water to wells that serve approximately 70% of the town's residents, as well as almost all residents of rural areas around the town.

- *Lake water* from Willow Lake, which supplies drinking water to approximately 30% of the residents within the town limits.

You will test these three water samples to see if they meet six of the federal water quality standards.

## Challenge

Working with a partner, you will conduct selected water quality tests to find out whether three water samples are suitable for their current uses. Be sure to keep careful records. Compare your results with the results of the other team at your table. If necessary, repeat tests to confirm your results.

# Water Quality

> **Federal Water Quality Standards to Be Tested**
>
> 1. **Appearance/Turbidity:** Water should be clear and free of any suspended materials.
> 2. **Odor:** Odor should not exceed a threshold value of 3—noticeable but acceptable.
> 3. **pH:** A pH level between 6.5 and 8.5 is acceptable.
> 4. **Salinity (salt):** Salt levels should not be greater than 250 ppm. Higher levels pose a health risk to some people with high blood pressure.
> 5. **Nitrate:** Nitrate levels should not exceed 10 ppm. Higher levels may cause serious problems, including brain damage or death, in infants. Nitrates prevent hemoglobin in the blood of infants and some sensitive adults from carrying enough oxygen through the body.
> 6. **Mercury:** Mercury should not exceed 0.002 ppm (2 ppb). It is concentrated in the body slowly over time and can lead to a variety of problems, including damage to the nervous system.

## Materials

*For each group of four students:*
- One 30-mL dropping bottle of each of the following:
  - Universal Indicator solution
  - Solution 2—0.1 M potassium chromate solution
  - Reagent A—0.1 M silver nitrate solution
  - 5% ammonia
- One 60-mL dropping bottle of 0.5 M hydrochloric acid (HCl) solution
- One container of Nitrate Indicator Powder
- One 20-mL graduated plastic tube with cap
- Two stirring sticks
- Two droppers
- Two SEPUP trays

Activity 20

# 20 Water Quality

## Procedure

Begin by observing the turbidity and odor of the three water samples from Silver Oaks, using the bottled water provided for comparison. Then follow the procedures below to test the pH, salinity, nitrate, and mercury levels of each of the water samples.

### I. Testing the pH

Add one drop of Universal Indicator to 10 drops of each sample in separate cups of the SEPUP tray. Use the chart below to find the pH of the sample.

**Safety**

Do not taste the water samples provided. Use caution when working with chemical solutions. Wear safety eyewear and avoid spilling the solutions on your skin. The silver nitrate solution will stain your skin and clothing. Be sure to wash your hands at the end of the activity.

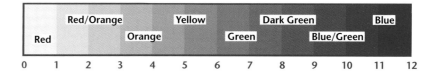

### II. Testing for Salinity (Salt)

1. Add one drop of 0.1 M potassium chromate solution (Solution 2) to 10 drops of each sample in separate cups of the SEPUP tray.

2. Add one drop of 0.1 M silver nitrate solution (Reagent A) to the sample and stir.

3. If, after stirring, the red color disappears, continue adding silver nitrate solution a drop at a time, and stir after each drop. Keep a record of how many drops of silver nitrate solution have been added. Do this until the red color does not disappear with stirring. Do not add more drops.

4. Use the standard graph on the next page to determine the salinity in ppm.

# Water Quality

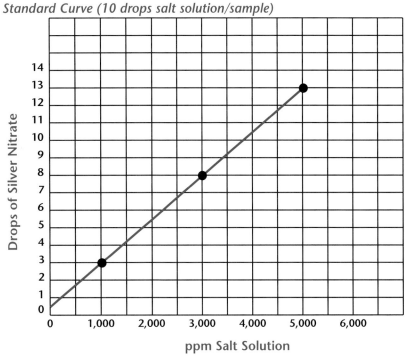

*Standard Curve (10 drops salt solution/sample)*

### III. Testing for Nitrates

1. Add 3 mL of your first sample (fill to the bottom small line) to the 20-mL graduated plastic tube with cap.
2. Add 10 drops of 0.5 M hydrochloric acid solution.
3. With your stirring stick, add 2 level measures of Nitrate Indicator Powder.
4. Replace the cover and shake until the powder is completely dissolved. Wait one minute.
5. A purple color indicates the presence of nitrates. (**Note**: The smallest concentration of nitrate this test can detect is 20 ppm.)
6. Repeat Steps 1–5 for each sample.

### IV. Testing for Mercury (Simulated Test)

1. Add 5 drops of 5% ammonia to 10 drops of each water sample in separate cups of the SEPUP tray. If a light blue color is observed, mercury levels are over 2 ppb. (This is a simulated test. Copper, a much less toxic heavy metal, has been used to simulate mercury.)

# Water Quality

 **Data Processing**

Prepare a full investigation report. Start your report on a clean sheet of paper. Include your name, the date, and a title for your report. Your report should have the following four components:

1. A statement of the problem you were trying to solve.
2. A description of the materials and procedure you used to solve the problem.
3. A clear presentation of the results you obtained.
4. An analysis of the results. You should include any calculations you performed and discuss any problems you had with the experiment or sources of error you may have had.

The analysis section of your report for this activity should include the following:

a. A discussion of how all water samples compare to the federal standards.

b. Your ideas about what could have caused any contamination you find in the water.

c. Your ideas about treatment of any of the samples to make them more suitable for use.

d. A summary of your recommendations to the town of Silver Oaks—tell whether each water sample is suitable for its current uses. Give reasons for your decision. Discuss any overall problems raised by your results and your suggestions of how the community should respond to your findings.

# Water Quality

## Introduction

### Nitrate: A Drinking Water Concern

Groundwater is the source of drinking water for nearly half the population of the United States. The purity of this water has become a health concern because a number of potentially toxic substances have been discovered in groundwater serving communities throughout the nation. Many of these compounds are naturally-occurring substances that have always been present in the environment. One of the most common examples is nitrate.

Although nitrate occurs naturally in drinking water, high levels in groundwater usually result from human activities. Overuse of chemical fertilizers and some wastes are sources of nitrogen-containing compounds which are converted to nitrates in the soil. Nitrates are extremely soluble in water and can move easily through soil into the drinking water supply.

High levels can build up over time as nitrates accumulate in the water, but even at elevated levels, they are not likely to be a health hazard for most adults. However, at certain concentrations, above 10 parts per million, nitrates can cause adverse health effects in very young infants and susceptible adults.

## Challenge

Identify the concerns about nitrates in drinking water.

Activity 20

# Water Quality

### Nitrate in the Environment

The most common sources of nitrate are municipal and industrial waste waters, refuse dumps, animal feed lots, and septic systems. Other sources are runoff or leachate from manured or fertilized agricultural lands and urban drainage. In addition, nitrogen compounds are emitted into the air by power plants and automobiles and are carried from the atmosphere to the ground with rainfall.

### Health Effects

When nitrate is ingested, bacteria in the mouth convert the nitrate in saliva into nitrite. The most significant health effect associated with nitrate ingestion is methemoglobinemia ("blue-baby syndrome") in infants under six months of age. This condition results from the presence of high nitrite levels in the blood. Methemoglobinemia occurs when hemoglobin, the oxygen carrying component of blood, is converted by nitrite to methemoglobin, which does not carry oxygen through the body as efficiently. As a result, vital tissues, including the brain, receive less oxygen than they need. Untreated, severe methemoglobinemia can result in brain damage and even death. In addition to small infants, some adults may be susceptible to the development of nitrite-induced methemoglobinemia. Fortunately, methemoglobinemia is easily recognized by the medical and public health communities and can be readily diagnosed and treated.

# Water Quality

## Nitrate as an Indicator

Nitrate in groundwater is of concern not only because of its toxic potential, but also because it may indicate contamination of the groundwater. If the source of contamination is animal waste or effluent from septic tanks, bacteria, viruses, and protozoa may also be present. Contamination of groundwater by fertilizers may also indicate the presence of other agricultural chemicals such as pesticides. The source of the nitrate may be a clue as to which other contaminants may be present.

## Remedies

Drinking water containing more nitrates than 10 ppm should not be consumed by infants or other susceptible individuals. Water that is bottled or taken from another safe source should be used. Simple in-line filters do not remove nitrates; but deionization, reverse osmosis, or distillation can be effective in the removal of nitrate. However, these treatments are expensive and require careful maintenance. In some cases drilling a deeper well extending into a noncontaminated water source may be the best and, in the long run, the least expensive remedy.

*By Mike Nugent and Michael Kamrin, Center for Environmental Toxicology; and Lois Wolfson and Frank M. D'Itri, Institute of Water Research, Michigan State University*

# Section D

## Investigating Groundwater
## Activities 21–28

### Overview

Over 75% of the people in the United States obtain their drinking water supply from water that travels beneath the Earth's surface. If you could dig a hole below the school grounds, you might hit water at 30, 60, or even 200 feet, depending on where you live in this country. Long ago this water fell as rain and soaked into the soil. Probably starting near mountains, or upstream of large rivers, the water has slowly traveled underground through layers of soil and rocks that act like sponges to soak it up. This water is called *groundwater*.

Silver Oaks is a town of 25,000 people, outside a large American city. It has a problem. As you discovered in the last activity, a well-water sample contains dangerous amounts of mercury—a heavy metal that can be stored in the body. Mercury can produce many serious chronic health effects. In this section, you will act as detectives to discover the source of the mercury contamination. After locating the source, you will propose some methods to clean it up. With your help, Silver Oaks can solve its water contamination problem.

# Activity 21

# Trouble in Silver Oaks

## Introduction

### The Silver Oaks Story

Most of the time we take a safe drinking water supply for granted. What happens when the drinking water supply is threatened by contamination?

## Challenge

Your challenge is to think about how you would react if you lived in Silver Oaks and your drinking water supply were threatened by contamination.

Silver Oaks is a town of 25,000 people in an area that used to be primarily agricultural.

# Trouble in Silver Oaks

Carla and her grandparents live in Silver Oaks. Carla likes living here because she can walk to the large community park and meet her friends. She learns about the town's history when she walks in the early evenings with her grandfather through the city cemetery to see the old tombstones near the lake. The town has changed dramatically since it was founded in the early 1800s. Areas that were once wooded are now used for factories, schools, businesses, subdivisions, and a landfill.

Willow Lake is in the northeastern section of Silver Oaks. The lake once provided the water necessary for many of the industries that operated in Silver Oaks. Now many of these plants have closed, leaving deserted factory buildings. The people who live in central Silver Oaks and the older homes in Golden Oaks use water piped to their houses from the Silver Oaks Municipal Water District. The water district uses water from wells and from Willow Lake. This water is treated to meet all federal standards before it is piped to the customers. The residents in outlying areas still use water from their own wells.

Carla lives in Silver Oaks Estates, a section of Silver Oaks that is not yet hooked up to the Silver Oaks Municipal Water District. Recently, Carla and her family have noticed a funny smell near their well and suspect that it may be contaminated. As a precaution, her family has started drinking bottled water. In addition, Carla's grandmother has noticed more outbreaks of "flu" in her neighborhood this year than in previous years.

**STOP** and discuss with your group important pieces of information, or evidence, that you have learned so far about the problem in Silver Oaks.

*Trouble in Silver Oaks*

# 21

Carla thinks that the problems people are having might have something to do with the water. She is studying water testing in her science class at school. She decides to collect water samples from her well, Willow Lake, and Fenton River. After students in her science class test the three samples of water, they find that there is a concentration of 3 ppb (parts per billion) mercury in the well water. Carla remembers from her study about cholera that a map might be helpful. As she begins to record the information on a map, she thinks to herself, "Here's my chance to be a detective!"

After completing the map, Carla and her grandparents talk with the health commissioners of Silver Oaks about the test findings. The health officials are impressed with the information that Carla collected and the map that she made. They are concerned about the source of contamination and how quickly it may be spreading through the underground water supply (*aquifer*). The health officials ask Carla and her classmates to help decide where wells should be drilled for testing.

### Question

Imagine that you live next to Carla and her grandparents. What would you do about the problem? Be sure to describe your concerns and any steps you would take after learning about the problem with the Silver Oaks water.

# 21

## Trouble in Silver Oaks

### Map of Silver Oaks

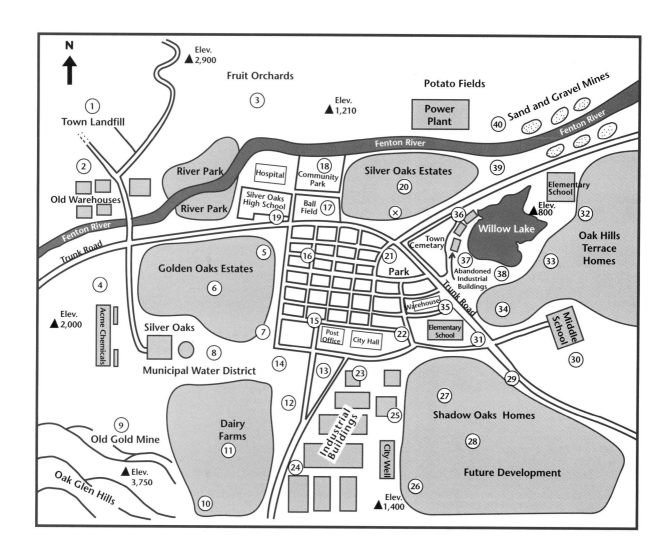

## Trouble in Silver Oaks

*Line Drawing of Silver Oaks*

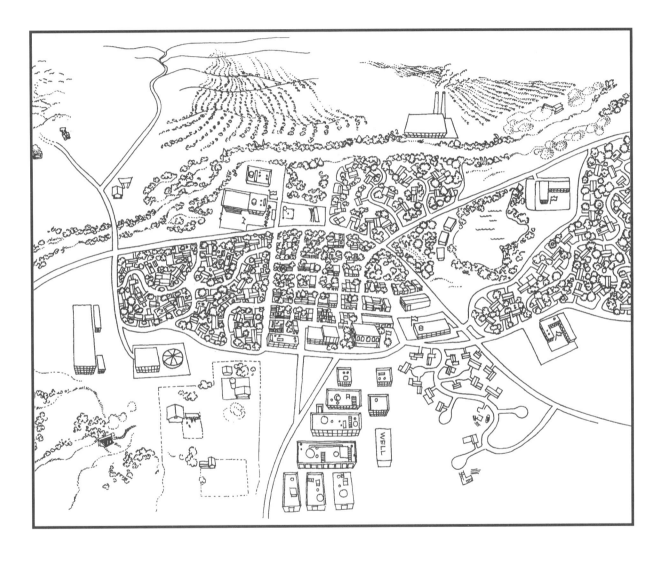

Activity 21

# 21

## Trouble in Silver Oaks

### Introduction — Earth's Groundwater Supply

You may think that most of our drinking water comes from sky blue lakes, crystal clear rivers, and large reservoirs. You would be wrong! These sources are only a drop in the bucket, compared to the abundant water supplies found underground. This water supply is called *groundwater*.

### Challenge

> Use the reading and diagram to learn more about groundwater.

Think of all of Earth's water. Imagine that all of it is contained in a 2-liter soft drink bottle. If you took away all the salt water represented in this bottle, the remaining water would represent fresh water and could be contained in a small glass (52 mL or approximately 2 ounces—the amount of water in a popsicle!). If you now took away all of the groundwater and water contained in ice caps and glaciers, you would be left with all of the Earth's water in lakes and rivers represented in our model by 0.28 mL, or less than 5 drops of water!

Groundwater is water that percolates down through the soil, like hot water through coffee grounds. It may fill up layers of soil or it may be stored between spaces in layers of rocks. These water-filled layers are called *aquifers* (full of water). As you can see, groundwater makes up a large part of our fresh water drinking supply compared to those sky blue lakes and crystal clear rivers!

More than 75% of the people in the United States obtain their drinking water supply from groundwater. Some of these people get the water from their own wells, while others obtain it from

*Trouble in Silver Oaks*

# 21

municipal (city or town) water companies. The rest of the country's water comes from surface water sources, such as lakes, rivers, and streams.

By its very nature groundwater is never "pure" water. Water from various sources, such as rain, snowfall, and agricultural irrigation, carries dissolved and suspended materials below the surface. It was once thought that soil could act as a natural filter and remove dangerous substances before they affected groundwater. Today, we find that our groundwater can be contaminated by such natural causes as radioactive minerals (radium) and gases (radon), toxic substances given off by plants and animals, and the action of microbes like giardia and amoebas. The human population adds to this contamination problem by the use of pesticides and fertilizers, leaking of underground storage tanks, poor sewage (septic tanks) and waste disposal practices, and air pollutants that dissolve in rain water.

Only 0.014% of the Earth's water is in streams, soil moisture, and lakes

2.6% of all water is FRESH WATER, including groundwater, ice caps, and glaciers

97.4% of all water on Earth is in the oceans

Groundwater and surface water (such as streams and rivers) are interconnected. Surface water helps to replace groundwater that has been removed for human use, while groundwater helps keep streams and rivers flowing when there is a drought. Groundwater makes possible a yearlong water supply, ensuring the health of lakes, rivers, streams, and oceans.

## Questions

1. What is groundwater? Why is it important?
2. Does the water you drink come entirely, in part, or not at all from groundwater?
3. How can surface water become contaminated? How can groundwater become contaminated?

# 21

*Trouble in Silver Oaks*

### Introduction

## Scoping It Out!

Silver Oaks has a water contamination problem. Sometimes it is helpful to organize information about a problem before continuing an investigation.

### Challenge

Use a table in your Journal to summarize the evidence and information about this mystery.

### Procedure

1. Prepare a table in your Journal like the sample below. Use a full page. Use what you have read about Silver Oaks to list *evidence* in the first column about the water contamination problem.

2. Use the second column to list any *ideas* you have that might help explain the mystery.

3. Use the third column to list *questions* you need to ask about water, mercury, Silver Oaks, or other information to help solve the contamination mystery.

| Evidence<br>What facts do I know? | Ideas<br>Ideas I have about the location | Questions<br>What do I still need to know? |
|---|---|---|
|  |  |  |
|  |  |  |
|  |  |  |
|  |  |  |
|  |  |  |

Activity 21

# Activity 22

# Water Movement Through Earth Materials

## Introduction

By investigating water movement through solid materials such as clay, sand, and gravel, you will begin to understand how mercury contamination may have traveled to the contaminated well in Silver Oaks.

## Challenge

Your challenge is to explain why water travels at different rates through different materials.

Tubes of sand, gravel, and clay can be used to model the flow of water through different earth materials.

# Water Movement Through Earth Materials

### Introductory Questions

1. List three examples of a liquid moving through a solid other than soil. Think of examples that occur in your daily life.

2. What determines how fast a liquid moves through a solid? Give two explanations.

3. Make a drawing that illustrates how a liquid travels through a solid.

### Procedure

1. Record the names of the materials in Tubes A, B, and C on a copy of the data table which appears on the opposite page.

2. Rank the materials from one to three (1–3) in the order you think they will transmit water. Use a number "1" for the fastest and a number "3" for the slowest. Record your predicted ranking on the data table.

3. Predict how long you think it will take for the water to reach the bottom of each tube. Record your predicted time on the data table.

4. Observe the demonstration of the water poured into each tube. Record on your data table the actual time it took for the water to reach the bottom.

5. Using the actual time results, complete the data table for the actual rank order of water flow through the earth materials.

# Water Movement Through Earth Materials

*Water Movement Through Earth Materials*

| Tube | Material | Predicted Rank | Predicted Time | Actual Rank | Actual Time |
|---|---|---|---|---|---|
| A | | | | | |
| B | | | | | |
| C | | | | | |

**Data Processing**

Explain why the water moved faster through some types of earth materials than through others.

# Activity 23: The "Ins and Outs" of Groundwater

## Introduction

### The Global Water Cycle

The world's freshwater supply is constantly recycled through the global water cycle.

## Challenge

Trace the path of fresh water as it moves through the water cycle.

All the fresh water in the world's lakes and creeks, streams and rivers, represents less than 0.01 percent (1 part in 10,000) of the Earth's total store of water. Fortunately, this freshwater supply is renewed by the precipitation of water vapor from the atmosphere as rain or snow.

The global water cycle has three major pathways: precipitation, evaporation, and vapor transport. Water precipitates from the sky as rain or snow. Most of this water falls into the oceans, but some of it falls on land to renew the freshwater supply. Some water flows from the land to the sea as runoff or groundwater. The water evaporated from the sea and from the surface water on land enters the atmosphere as water vapor. The water vapor is then transported over the land by atmospheric currents. The cycle begins again as the water precipitates from the sky to the land.

*The "Ins and Outs" of Groundwater*

# 23

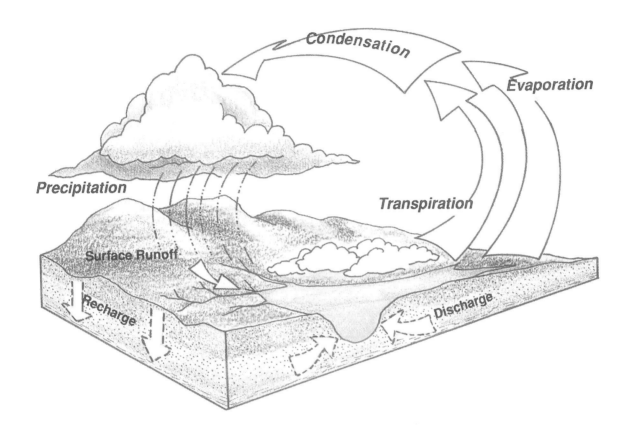

When surface water sinks into the ground, it renews or recharges the groundwater supply. This process can also go in the opposite direction. Groundwater is discharged to the surface at natural springs.

As water moves through the water cycle, particles and dissolved materials can be transported. When the water evaporates, it enters the atmosphere as pure water vapor. In the atmosphere, it can become contaminated by gases and particles that result from both human activity and natural events such as volcanoes. As water runs over the land, it picks up additional materials that are eventually carried to rivers, lakes, and the ocean.

# 23

## The "Ins and Outs" of Groundwater

### Introduction

**Understanding Groundwater**

The ground below your feet can act as a giant sponge. It can absorb the rain and transport it deep underground where it may encounter layers of rocks that can absorb it or hold it back from moving to deeper layers. Eventually all water traveling underground returns to the ocean. How fast does this process happen?

### Challenge

Does water travel at the same speed through different types of soil and rocks? In this activity you will explore the answer to this question and discover how earth materials act as *aquifers* and *aquitards*.

# The "Ins and Outs" of Groundwater

## Materials

*For each group of four students:*
- One 9-ounce plastic cup filled with water
- One small plate or dish for rock samples
- 5-cm square samples each of granite, limestone, and sandstone
- One dropper
- One 25-cm x 3.75-cm plastic tube with holder
- One plastic vial with lid
- Four soil samples from home
- One graduated cylinder

*For each pair of students:*
- One hand lens

## Procedure

### Part One: Separating Soil Particles

1. Fill a plastic vial half full with a "Standard Soil Sample."
2. Add water to the sample until the water level is near the top of the vial.
3. Cap and shake the vial.
4. Allow the contents to settle out.

### Data Processing

1. Draw a picture of the layers formed in the sample. Label them.
2. Why do you think the layers are in this order? Explain.

*Continued on next page* →

# The "Ins and Outs" of Groundwater

**Procedure (continued)**

### Part Two: Testing Soil Samples

1. Examine your soil sample with a hand lens. Describe what you see in your data table. Use a table like the one below.

2. Fill the large plastic tube with your soil sample, and tamp it down lightly so that the soil settles.

3. Predict how long it will take for water to travel to the bottom of the container. Base your prediction on the observations you made in the last activity comparing samples of gravel, sand, and clay, as well as your observations of what your sample contains. Record your predicted time.

4. Pour 30 mL of water into the top of the tube and record the amount of time it takes for water to first reach the bottom of the tube. Record your actual time.

*Observations of Soil Samples*

| Sample | Description of Sample | Predicted Time | Actual Time |
|---|---|---|---|
| 1 | | | |
| 2 | | | |
| 3 | | | |
| 4 | | | |

### Data Processing

1. Is your sample of soil more like the sand, gravel, or clay that you observed in the last activity? Or is it a combination of all three?

2. Observe the results of other groups and write a short paragraph in your Journal comparing their samples with yours.

Activity 23

# The "Ins and Outs" of Groundwater

## Part Three: Testing Rock Samples for Permeability

1. Identify the three rock samples at your table. Use the hand lens and record your observations in a table like the one shown below.

**Safety**
Do not chip or break rocks. Some rocks may have sharp edges.

2. Add one drop of water to the surface of each rock sample and record your results in the data table.

3. Use a scale from one to three (1–3) to rate how quickly water flows into each rock. Let "1" represent very slow water flow, and "3" very fast water flow into the rock.

*Observation and Testing of Common Rocks*

| Rock | Origin | Description | Permeability (Use scale 1-3) |
|---|---|---|---|
| granite | deep underground from cooling magma (hot, liquid rock) | | |
| sandstone | from layers of sand deposited in water | | |
| limestone | from layers of shells deposited in water | | |

## Data Processing

1. Which rocks were most permeable to water? Which rocks were least permeable? What is your evidence?

# The "Ins and Outs" of Groundwater

*Part Four: Groundwater Movement*

Surface water trickles down through the Earth at a rate of several inches to several feet per day. This water may reach layers of rock like granite, sandstone, or limestone. In permeable rocks, the water collects like a sponge full of water at various depths below the surface. This is groundwater. The diagram below shows the movement of groundwater and some of its sources. The arrows represent the direction of water movement.

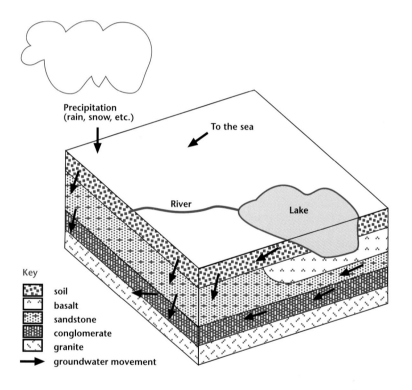

## Questions

1. What is an aquifer? List the aquifers on the diagram by using the name of the layer of material they are made of.

2. What is an aquitard? List the aquitards on the diagram by using the name of the layer of material they are made of.

3. Describe the flow of the water from the surface of the Earth through the aquifers to the conglomerate level.

# The "Ins and Outs" of Groundwater

## Introduction

### Mercury

Of all the elements known to ancient people, mercury was regarded as having magical properties because it was the only metal known to exist as a liquid at room temperature. This reading will help you understand some of the interesting properties of mercury, and how it can be highly toxic to humans.

## Challenge

Use this reading to learn more about how mercury may have contaminated the Silver Oaks groundwater and about the health risks of mercury.

Also called quicksilver, mercury is one of only three liquid elements. Like gold, silver, and copper, it is one of the few elements that is found in nature in its pure form.

Ancient people believed mercury had magical healing powers and used it to cure diseases. Not long ago, mercury was made into a compound called mercurochrome. It was used to kill bacteria in cuts. OUCH! Until recently, seeds were coated with mercury compounds to prevent them from molding when placed in damp soil. Mercury was also used to process the felt that was placed in hats. People eventually learned of its highly toxic properties. The expression "mad as a hatter" comes from the observation that people who used mercury compounds to make hats showed symptoms of mental illness!

*Continued on next page →*

# The "Ins and Outs" of Groundwater

Mercury has many uses today. One early use was in thermometers because mercury expands uniformly when heated. It is used in barometers to measure air pressure. Mercury is used in the manufacture of plastics and pesticides. In your home, mercury is found in fluorescent tubes used for lighting. It is used in some thermostats, where it acts as a conductor of electricity for a switch that senses temperature changes.

Gold dissolves in mercury, forming what is called an amalgam. When mercury is placed with gold-containing ore, the gold dissolves in the mercury, then the mercury is boiled away, leaving the gold behind. Dentists use a mixture of silver and tin, dissolved in mercury, for fillings.

Mercury, as a metal, can produce both acute and chronic toxicity. Mercury compounds can be transported from mines, landfills, and incinerator gases into the environment. At very low doses the vapors can be inhaled and stored in the body, resulting in chronic mercury poisoning. As a compound, mercury chloride is quickly absorbed into the body. Very small amounts can result in an acute toxic dose and death.

When mercury is released into the atmosphere, it may also enter lakes and rivers. There bacteria transform it into a toxic form called *methylmercury*. Small microscopic plants and animals pick up methylmercury if it is in the water as they feed. Small fish eat these plants and animals in large numbers. If there is mercury in the water, the concentration may increase in these fish from parts per trillion to parts per billion. These fish may then be eaten by larger fish, such as pike, tuna, and swordfish. In these large fish the mercury may now be hundreds of parts per billion. The mercury increases, or magnifies, as it travels through this food chain. If you regularly eat fish that have accumulated high concentrations of mercury, your chances of mercury poisoning are greatly increased.

# The "Ins and Outs" of Groundwater

However, most people do not eat enough fish for the mercury levels in their bodies to rise dangerously high. Therefore, the average person is at no great risk from exposure to mercury. But in countries where fish intake is high and there is significant mercury contamination, or in areas of high local pollution, the population is at risk for mercury poisoning.

The source of nearly 50% of the methylmercury in the environment is combustion. Industrial plants and waste incinerators that process household wastes containing mercury (batteries, electrical switches, and fluorescent lights) emit a small, but significant, amount of mercury gas into the atmosphere. This gas dissolves in rain and is transported to bodies of water. Over half of the mercury contamination in a Canadian lake was traced to the atmosphere, and the other half came from natural mineral sources. Acid rain provides a double whammy in transporting the mercury vapor and then providing an acid environment friendly for bacteria to form methylmercury.

## Questions

1. Now that you have read about mercury, what concerns do you have about the contaminated water that was found in the Silver Oaks well?

2. Based on the reading, where might the mercury in the well water in Silver Oaks have come from?

# Activity 24

# Investigating Contamination Plumes

## Introduction

### Modeling Contamination Plumes

Toxic substances can contaminate an area when they leak or are spilled from a very concentrated source of material. This happens when a barrel or an underground storage tank leaks. These are called *point sources* of contamination. When a toxic material is spread over a large area, such as when pesticide is sprayed from an airplane, the source is called a *non-point source* of contamination.

## Challenge

Your challenge is to determine how point and non-point sources of contamination change the shape of an underground plume. You will investigate two models of plumes to find out. The sand in this activity represents an aquifer and the food coloring represents a contaminant, such as mercury.

## Materials

*For each group of four students:*

- One 30-mL dropping bottle of red food coloring
- One plastic cup of water
- One small graduated cup
- One waste container

*For each group of two students:*

- One large (100-mm) plastic petri dish with sand
- One dropper

# Investigating Contamination Plumes

## Procedure

1. Each group of four students will divide into two pairs. Each pair will need a petri dish with sand. In your Journal carefully note observations and include drawings of what you see happening. You are also responsible for recording the other pair's observations in your Journal.

2. One pair will do Part One, and the other pair will do Part Two of the investigation.

### Part One: Point Source Contamination

*One pair in your group does this part.*

**Safety**

Food coloring can stain clothes and work surfaces. Immediately clean up any spills with water.

1. Use the dropper to slowly add water to the sand in the petri dish until all of the sand is damp, from top to bottom. If you add too much water, carefully pour it off into the dishpan provided by your teacher.

2. Lean one edge of your petri dish against your closed Student Book so that the dish is slanted. It should be supported by the book and not fall away from it.

3. Make a small hole in the sand in the middle of the dish. Add two drops of food coloring to the hole and cover it gently with sand. The drops of food coloring represent mercury.

4. Slowly add 20 drops of water to the sand at the top end of the petri dish and carefully observe what happens for the next 2 minutes.

5. Thoroughly examine the bottom of the dish and record all of your observations in your Journal.

6. Clean up as directed. Do not discard sand in the sink. Use the container provided!

Activity 24     143

# Investigating Contamination Plumes

## Procedure

*Part Two: Non-point Source Contamination*

*The other pair in the group does this part.*

**Safety**

Food coloring can stain clothes and work surfaces. Immediately clean up any spills with water.

1. Use the dropper to slowly add water to the sand in the petri dish until all of the sand is just damp, from top to bottom. If you add too much water, carefully pour it off into the dishpan provided by your teacher.

2. Lean one edge of your petri dish against your closed Student Book so that the dish is slanted. The dish should be supported by the book and not fall away from it.

3. Add a drop of food coloring to 15 drops of water in the graduated cup. Mix. This solution represents a more diluted form of mercury that will spread over a wide area.

4. Place four small holes in the sand across the middle of the dish. Place two drops of the diluted food coloring in each hole. Gently cover the holes with sand.

5. Slowly add 30 drops of water in a line back and forth across the top of the petri dish.

6. Carefully examine the bottom of the dish and record all of your observations in your Journal.

7. Clean up as directed. Do not discard sand in the sink. Use the container provided!

## Data Processing

Summarize the similarities and differences in how contamination from point sources and non-point sources moves through aquifers.

Activity 24

# Investigating Contamination Plumes

## Facts About Groundwater

1. Approximately 75% of the United States population depends, at least partially, on groundwater for drinking water.

2. An estimated 95% of rural households depend exclusively upon groundwater for their drinking water.

3. The federal government estimates that 2% of the groundwater is polluted from point sources. This is actually more of a threat than the low figure might indicate since the pollution tends to be near high population areas and, thus, affects more people. Also, this figure does not include non-point sources of pollution such as agricultural and urban runoff.

4. It is not uncommon for water from wells to exceed state maximum recommended limits for various substances. In 1984–85, for example, one-fifth of all large wells supplying drinking water in California exceeded state pollution standards.

5. Infections from pathogenic bacteria, viruses, and parasites cause about 50 times as many cases of acute illness as chemical contamination of water.

6. Apart from microbes, virtually every contaminant causes an acute response only if the substance is present in large amounts. In the quantities in which most contaminants are present, the main risk is of chronic, low-grade illnesses which are hard to diagnose.

# 24

## Investigating Contamination Plumes

### Introduction

**Developing Your Well Testing Plan**

In this activity and the next activity you will help the people of Silver Oaks. The community wants to determine the source of the mercury contamination and to measure how far it has spread. The first step is to drill three test wells and to test them to find out how much mercury is in the groundwater.

### Challenge

Your challenge is to decide on three locations where the water should be tested first.

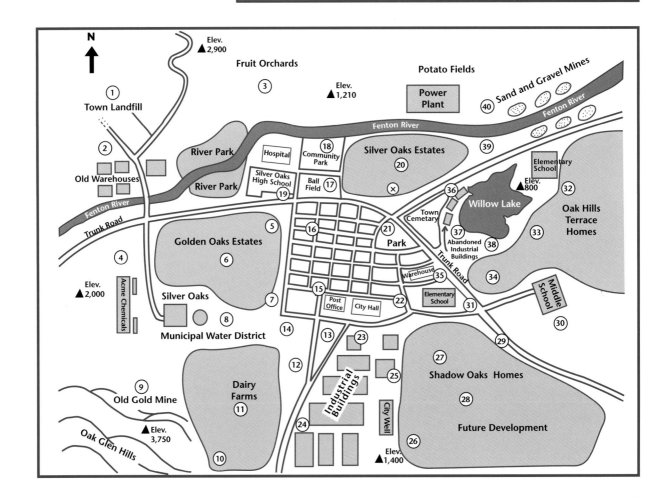

# Investigating Contamination Plumes

## Procedure

1. You know that the well marked with an "X" on the map is contaminated with mercury levels above the federal standards. The health commissioners of Silver Oaks have identified 40 possible sites for drilling test wells. However, the community only has enough money to drill 12 test wells. Using clues from the Silver Oaks story, information from the Silver Oaks map, and your notes on groundwater movement, decide which three wells you think would be most important to test first.

2. Record the three wells you have chosen and your reasons for choosing them in your Journal. Be prepared to explain your choices.

# Activity 25

## Testing the Waters

### Introduction

**Well Selection and Testing**

In the previous activity you learned about the movement of contamination plumes. You used this information, along with information about Silver Oaks, to choose the first three locations where wells should be drilled and tested. In this simulation, you will test well-water samples to determine the source of the mercury contamination and to measure how far it has spread. Remember, Silver Oaks' funds are limited—only 12 of the possible 40 well sites can be tested. You will test three wells at a time, and use the evidence you obtain from each phase of testing to plan the next phase.

### Challenge

Working as a group, you must test 12 wells, three wells at a time, to determine the source of the mercury contamination and to find out how far it has spread through the aquifer below Silver Oaks. You will use this information to develop an "underground" map of the contamination plume.

Drilling a well.

# Testing the Waters

## Materials

For each group of four students:
- One Chemplate—small plastic tray with 12 wells
- One 30-mL dropping bottle of Universal Indicator solution
- Paper towels or sponge
- Student Sheet 25.1, "Well Selection and Testing Report Form"
- Student Sheet 25.2, "Map of Silver Oaks"

## Procedure

1. Take out your Journal and turn to the page where you recorded your ideas about which three wells should be tested first. As a group, discuss the choices each of you have made. When you come to a consensus about which three wells to test first, record the three wells you chose and your reasons for choosing them in the Well Selection Table on the form on Student Sheet 25.1. Be sure to explain your reasons for your choices thoroughly.

2. Test the three wells for mercury. Each well is tested by adding 5 drops of the well water to one drop of Universal Indicator in the Chemplate.

3. Record the results of your test in the Well Testing Results Table. The chart at the bottom of this page tells you how to use the colors obtained with the Universal Indicator to determine the concentration range of the mercury and the code of the well.

4. You will continue to test the wells in groups of three until you have tested 12 wells and completed the report form.

**Safety**

Wear safety eyewear and avoid skin contact with the chemicals. The test used is a simulation. There is no mercury in the samples.

*Concentration Range Chart for Well Samples*

| Color | Concentration Range | Code |
|---|---|---|
| yellow-orange | not detected—less than 0.1 ppb | 1 |
| yellow | 0.11 ppb – 0.8 ppb | 2 |
| green | 0.81 ppb – 4 ppb | 3 |
| blue-green | 4.1 ppb – 32 ppb | 4 |
| blue | more than 32 ppb | 5 |

# ACTIVITY 26

## Mapping It Out

### Introduction

**Movement of Plumes**

The two maps below show the contamination distributions of the same area drawn by two different people. These two people tested different wells. They have marked the distance of the safe boundary from the source of the contamination. They have also marked the distance from the boundary to the city water wells. They know that the source area was contaminated only one time, and this contamination occurred five years ago.

### Challenge

Find out how quickly the contamination will reach the city water well.

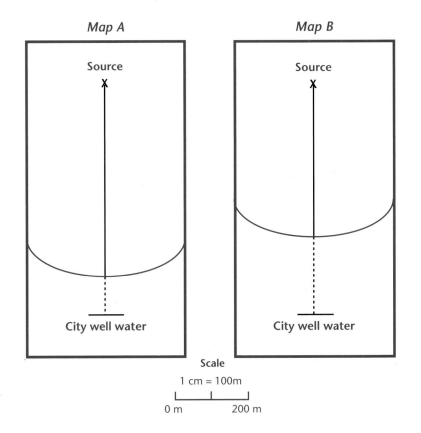

Activity 26

# Mapping It Out

**Questions**

1. Look at Map A. How far has the pollutant traveled in 5 years? How far has it traveled (on average) per year?

2. Look at Map B. How far has the pollutant traveled in 5 years? How far has it traveled (on average) per year?

3. If the pollutant continues to travel at the same average speed, how long will it take to reach the city wells?

   Map A:  It has to travel ___ meters at a rate of _____ m/yr. It will take _____ years.

   Map B:  It has to travel ___ meters at a rate of _____ m/yr. It will take _____ years.

4. Explain how two people's data could lead to such different predictions for when the contamination plume will reach the wells. Try to think of more than one possible explanation.

5. If you were a health officer for the city and these were the only plume maps available, which map would you use to make decisions about using water from the city wells? Explain.

# 26

*Mapping It Out*

## Introduction | Clean-Up Methods

There are several methods available for cleaning up contaminated groundwater.

## Challenge

Read about possible methods for cleaning up the groundwater contamination in Silver Oaks. Think about which method might work best. In Activity 28 you will have a town meeting to select a method.

### Methods Proposed to Clean Up Contamination

#### Containment

Surround and cover the area with clay. Drill test wells and install pumps at selected points in the path of the plume. Monitor the mercury concentration in the groundwater so that the wells can be shut down if the levels are too high. This method is quick and relatively inexpensive, minimizes air pollution, has been used for this purpose in other areas, and poses little danger to the public. Disadvantages include the need to monitor the water as long as the wells are used, issues of worker safety as the clay cap is installed, problems siting the clay layer, and the danger of well pumps actually pulling contaminated water from other areas into the aquifers as water is pumped out of them.

#### Removal

Use a pump to remove the contaminated water from the aquifer for treatment. This can be done quickly and has been used in other areas with relatively little danger to the public. However, pumping will not immediately remove contaminants trapped in sediments. Also, the treated residue may be classified as a hazardous waste and will require additional procedures for disposal.

# Mapping It Out

## 26

### Excavation

Dig out the earth materials in the plume area and transport the materials to a hazardous waste site for disposal. This process has been used successfully in other areas and does remove material that is highly hazardous. However, this process involves a high initial cost and the possibility of leakage and spills during excavation and transportation. In addition, it is often difficult to determine the exact extent of the contamination and to excavate in particular locations, such as heavily populated areas or under buildings. The exposure of the contaminant to the air may lead to other problems, and the pumping may redistribute the contamination in the aquifers, so it is necessary to monitor the water continuously with test wells.

### Excavation and Treatment

Dig out the contaminated earth materials and treat them with a new chemical process that uses a combination of water mixed with a mineral salt to extract the mercury compounds. The water evaporates. The remaining concentrated mercury compounds mix with clay. The clay mixture is placed in a high temperature kiln where the mercury combines with the clay and is held tightly to it. The rock-like product does not dissolve in water and does not form harmful leachates. These waste "rocks" could then be used as materials to make roads since the mercury would no longer be able to dissolve in water and be released into the environment. However, there is a high initial cost to construct the facility. There is also some danger of mercury vapors being released into the air during the heating process.

### Electrification

Use an underground electric field to decontaminate the groundwater. This is less expensive than other options and is effective for the entire area even though the exact boundaries are not well known. It is quick and has been safe in all tests to date.

*Continued on next page* →

However, as it has only been tested in the laboratory, there is some concern that it might not work in the field. The direction and extent of the electric field may not be as precisely controlled in the field as it has been in the lab.

### Shut Down the Water Source

If alternate sources of water cannot be located in the area, the residents of Silver Oaks may have to locate water from another source, possibly involving more cost and certainly involving inconvenience. An alternate source of water may not be readily available. It does not clean up the contaminated groundwater, and will not protect the plants and the wildlife from the effects of the mercury.

### Questions

1. What method(s) do you favor? Why?
2. What method(s) are you opposed to? Why?

# Activity 27

## Practicing Presentation Skills

### Introduction

Science provides evidence, but people make decisions. In this activity, you will practice your presentation skills to prepare for the public meeting in the next activity.

### Challenge

Use the following tips to help you play your role in a class meeting.

## Practicing Presentation Skills

### Role-Playing Tips

1. You are playing the part of a person involved in this issue. How are you involved and what might your opinion be?
2. What reasons might you have for this opinion?
3. Who might not agree with you? (Consider the other roles.)
4. What reasons might be given for the disagreement?
5. How would you respond to opposing arguments?
6. List the key points that you want to make clear in your presentation to the audience.
7. How would you behave at a public meeting? How would you dress? How would you greet people? How would you talk?

You may want to bring props and dress in costume for your part. *It is strongly suggested that you practice your presentation in advance.*

### Audience Participation Tips

1. You are playing the role of a member of the audience. How does the issue affect you?
2. Have you already decided what you think should be done?
3. Can someone convince you to change your opinion?
4. What questions do you want answered at the meeting?
5. How will you dress for the meeting?
6. How will you behave at the meeting?

During the meeting, you will be given an index card on which to write questions for members of the panel. The questions can be those you thought of earlier or any that may occur to you during the meeting. Direct your questions to specific members of the panel. Be sure that you get all the information you need to make a thoughtful decision.

# Activity 28

## Cleaning It Up

### Introduction

**The Town Meeting**

You have mapped the underground contamination plume and know where the source is located. Now it is time for the community to take action.

### Challenge

Decide on the best method for cleaning up the mercury contamination in Silver Oaks.

---

## Silver Oaks Beacon

**Threat to Planned Shopping Mall**

The recent discovery of mercury contamination on the proposed site of the new Silver Oaks shopping mall has halted progress on approval of the site. Local laws do not allow the land to be developed until it is cleaned up. Until these recent developments, the site appeared to have the go-ahead, with final approval by City Council scheduled for next month. The downtown shopping area can no longer serve the growing population of Silver Oaks and the surrounding area. Citizens surveyed by this reporter were disappointed with the new turn of events. "I was so excited, but now I'm worried that the new mall will never be built," said one young shopper.

---

### Procedure

Hold a town meeting to decide on the best clean-up plan for Silver Oaks. Remember to use the Role-Playing Tips and Audience Participation Tips from Activity 27. Do your best to act your assigned role. Role players: Don't forget the persuasive power of evidence as you prepare your presentation!

# Part Two

# Materials Science

# Part 2

## Table of Contents

*Introduction* ..........................................................................................................1

29. Differentiating Humans from Other Organisms ..................................3

30. Materials Through Time ..............................................................................8

31. Properties of Materials ...............................................................................17

32. Conductors and Insulators ........................................................................24

33. Preventing Corrosion .................................................................................29

34. Bag It! Paper or Plastic? .............................................................................42

35. Properties of Plastics ..................................................................................46

36. Synthesizing Polymers ...............................................................................57

37. Paper Clip Polymers ...................................................................................63

38. Which Packing for Mike's Games? ...........................................................73

39. Comparing Garbage ...................................................................................79

40. Investigating Sanitary Landfills ................................................................82

41. Products of Hazardous Waste Incineration ...........................................91

42. Recycling Materials .....................................................................................98

43. Chemical Change: The Aluminum-Copper Chloride Reaction ........107

44. Reducing and Recycling Hazardous Materials ....................................118

45. Source Reduction .....................................................................................123

46. Integrated Waste Management .............................................................137

# Part 2: Materials Science

## Introduction

You will learn about materials science in Part Two of *Issues, Evidence and You*. Materials science deals with the selection, production, and processing of materials for a wide range of uses. Advances in civilizations have taken place because people have learned to craft new materials or to use materials in new combinations and designs.

Materials may be used in many places—homes, schools, businesses, factories, or automobiles. In the following activities, you will learn how scientists and engineers select materials for specific purposes. You will also learn about plastics, the newest materials produced since the discovery of metal alloys. Plastics are designer materials. They can be created and designed to be used for specific purposes, such as super glues or even artificial heart valves.

The materials we create can eventually cause problems. When their useful life ends, they become wastes that are discarded into the Earth's air, water, and land. In this section, you will study about the problems of these "leftovers" that we call garbage. You will learn about something called *life cycle analysis,* a method you can use to analyze the impact any consumer product you buy, use, and throw "away" has on your environment.

Materials touch your life in many ways. Just about everything in your world is a result of materials science—from the silver fillings in your teeth to the plastic pens you use for writing to the personal stereos, microwaves, or other electronic products in your home. Let's turn the page and take a closer look at the world of materials, the world Lewis Carroll described in *Through the Looking Glass*, "The time has come, the Walrus said, to speak of many things: of shoes and ships and sealing wax, of cabbages and kings..."

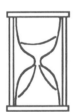

### Keeping Track of Time

America and other Western countries use the Gregorian calendar to keep track of time. This system is named after Pope Gregory XIII who developed it in the sixteenth century.

"A.D." stands for "anno Domini" or "in the year of the Lord." For Christians, the first century of our system begins in A.D. 1, the year Jesus was born. A.D. is also called "C.E." which stands for "of the common era." C.E. is used especially by non-Christians.

The centuries are counted as follows:
 A.D. 1–A.D. 100 = first century
 A.D. 101–A.D. 200 = second century

Can you see why we call the current century (A.D. 1901–A.D. 2000) the 20th century?

The year before A.D. 1 is not called A.D. 0, but instead is designated 1 B.C. ("before Christ") or 1 B.C.E. ("before the common era"). This sequence is shown on the timeline below.

**3 B.C.  2 B.C.  1 B.C.  A.D. 1  A.D. 2  A.D. 3**

Other calendars, such as the Chinese, Muslim, and Hebrew calendars, are also used throughout the world. How do the years from these calendars correspond to our system?

In A.D. 2000 it will be:

| Year | Calendar |
|---|---|
| 1362 | Burmese Thingyan Tet |
| 17 | Chinese (78th Cycle) |
| 1994 | Ethiopian |
| 5760 | Hebrew |
| 1922 | Saka (Indian) |
| 2390 | Zoroasterian |

# Activity 29

## Differentiating Humans from Other Organisms

### Introduction

**Hand Craft**

Throwing stuff away is an ancient—and still common—human activity. Is it something only humans do, or do other kinds of living things also throw away unwanted stuff? Maybe it's the throw itself that contains some clues about this behavior. Picking up something, holding it, and then doing something with it (like throwing it) seems so natural that we hardly give the task a thought. But what would tasks that require the use of your hand be like without your thumb?

### Challenge

Try to carry out a variety of activities without using your thumb.

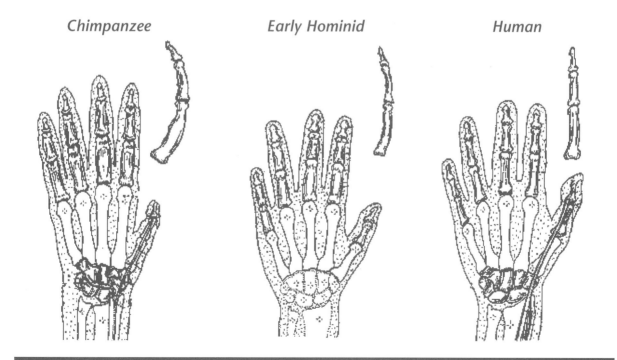

*Chimpanzee*  *Early Hominid*  *Human*

# 29 Differentiating Humans from Other Organisms

## Materials

*For each pair of students:*
- One zip-lock bag
- Five paper clips
- One nut and bolt
- One timer (or student watches)

## Procedure

You will pit your manipulative skills against those of your partner and then your team's skills against those of your classmates by playing a game called "Terminal Phalanx." The basic idea of the game is simple. Each team is timed on how long it takes both members to complete a series of tasks both with and without the use of their thumbs. The team with the shortest combined time is the winner. Before beginning, decide which team member will do the tasks first and which one will be the observer. (You will be doing both so it doesn't really matter.) When you are the observer, you must time your opponent and make sure at the appropriate time that he or she is not using either thumb. *Make a data table in your Journal to record times.*

# Differentiating Humans from Other Organisms

## The Tasks

- Unzip the zip-lock bag and remove the contents.
- Make a chain with the 5 paper clips by linking them together.
- Thread the nut completely onto the bolt—all the way to the end.
- Put the chain and threaded nut into the bag and tightly zip it.

## The Rules

- Play begins with the sealed zip-lock bags on the table in front of each opposing player. Each bag must contain 5 paper clips (unlinked) and a nut and bolt (unthreaded). You may not touch the bag until time is started.
- **Round one is with the use of thumbs.** Be sure that you switch roles and reset the zip-lock bag so that everyone gets a recorded time.
- **Round two is without the use of thumbs.** When you are the player, you must tuck your thumbs into the palms of your hands and keep them there. If the observer notices that the thumb was used, the player must start that individual task over again. Switch roles.
- Record the times for you and your partner for each round. Compute the total times for your team.

# 29 Differentiating Humans from Other Organisms

## Introduction

### In the Company of Others

Many human behaviors, including uses of material, are also demonstrated to some extent by other living organisms. Maybe early people even learned from watching animals!

## Challenge

After completing this reading, think of other examples of complex human activities carried out by animals.

Although we share the planet with many other kinds of living things, it is easy for us to feel different and unique. Sometimes we describe our way of life—the things we do and the way we do them—as the "Human Condition." But can we agree on what is human, and only human? Most likely you can think of other living organisms that learn, play, fight, have feelings, and communicate. For example, scientists have observed chimpanzees laughing as they tickle or play tricks on one another. Other animals perform elaborate courtship rituals to communicate. The courtship display and ritual dance of male and female cranes is their way of overcoming their natural fear of one another and forming a strong bond. Once a pair mates, they stay partners for life.

Human family relationships aren't unique either. Having parents (or sometimes adoptive parents) that care for their young is also common among other animals. One example is ostrich families. They set up daycare. After the young ostriches hatch, they gather together in multi-family groups, which are looked after by just one adult.

# Differentiating Humans from Other Organisms

We aren't even alone in the basic ways that we get our food. Some animals "farm" or tend "livestock." In South America there is a type of ant that "farms." The ants work together carrying pieces of leaves to their underground nest where they use the leaf pieces as fungus garden beds. The ants "sow" the beds with the spores of a particular fungus that grows quickly. These garden beds can be several feet in length. Another type of ant raises aphids. These ants will even fight to protect the aphids from their enemies. The ant colony keeps the aphids near a good supply of food. In exchange for looking after the needs of the aphids, the ants "milk" the aphids for the sweet nectar they produce. They do this by gently stroking their backsides. Doesn't this seem a lot like dairy farming?

For a long time humans thought that one trait that distinguished them from other living organisms was the ability to use tools to make life easier. But humans are not the only tool-users. The Egyptian vulture uses stones to break open the thick, hard shell of ostrich eggs. The vulture picks up a stone in its beak and throws it at the egg. Now we know why ostriches set up daycare programs! Other birds use plant parts as tools. The woodpecker finch, a bird that lives on the Galapagos Islands, uses cactus spines to pick out insects and grubs from the bark of trees.

Some animals actually adorn themselves with "clothes." The caddisfly larva, which lives in water, weaves itself a silk tube-shaped covering and then sticks bits of gravel, twigs, and leaves to it. What a fashion statement!

### Question

Can you think of any human activities not done by animals? Just what is it that sets us apart from other animals?

# Activity 30
## Materials Through Time

### Introduction

**Human Use of Materials**

Look at the objects on these two pages. Some are hundreds or thousands of years old, while others are from our own time. These items are made of different materials.

# Materials Through Time

## Challenge

Observe each object carefully and try to determine what materials it contains. Using the number of each picture, work with your group to put the pictures in order by age of the object with the oldest first.

7

8

9

11

10

12

Activity 30

# Materials Through Time

## Introduction — Garbology Survey

In order to deal effectively with garbage problems, it is important to know what the problems are and which problems the public thinks are important. Since you and your classmates are members of the public, the following survey of your ideas will provide some evidence about what one group of the public thinks.

## Challenge

> Answer the survey question in your Journal. Be sure to give your reasons for your answer. Do you think you have enough evidence to make a decision?

Most of the studies of garbage indicate the following are the contributors to garbage problems. Which one do you think is the greatest cause of garbage problems? Rank these, starting with the one that you think takes up the most space (volume) in landfills.

a. Disposable diapers

b. Food and yard waste

c. Newspapers

d. Large appliances

e. Construction debris

f. All paper

g. Plastic bottles

# Materials Through Time

### Introduction

## The Great Garbage Mystery

Studying garbage can tell us many things about garbage and people. In this reading, two students survey their community and get some results that conflict with their observations.

### Challenge

Consider the value and limitations of using surveys as a method to obtain evidence.

Willis and Darnell live in a big city in the northeastern United States. The neighborhood around their school has a problem with garbage. There is a lot of trash on the streets, in vacant lots, and around the school's athletic field. A lot of the trash is cans, bottles, and paper. All of these items can be recycled. Every week in their neighborhood, the city's recycling truck picks up cans, bottles, and paper for recycling.

For science class, Willis and Darnell decide to survey people who live in the area around their school. The survey asks how often people recycle their cans, bottles, and paper. The results of the survey are surprising! Almost everyone who was surveyed says he or she always recycles paper, cans, and bottles.

Seeing all the cans, bottles, and paper trash in their neighborhood, Willis and Darnell wonder whether there is something wrong with their survey. Most people who live in the neighborhood say they recycle these items. So why, the students ask themselves, are there still a lot of cans, bottles, and paper trash scattered around the neighborhood?

 **Question**

What could explain the difference between the students' survey results and the amount of trash that they see in their neighborhood? Try to think of two or three possible explanations.

Activity 30

# 30

## Materials Through Time

### Introduction

**Using Archaeology to Learn About Today**

This reading will help you learn more about The Garbage Project and the science of garbology, including its relationship to archaeology.

### Challenge

> If researchers from The Garbage Project asked you to tell them what was in your garbage last week, would they find the same discrepancies between your descriptions and the evidence as the discrepancies described in "The Great Garbage Mystery"?

Archaeology tells us many things about the past. It can tell us how people behaved, what they believed, and what they did in their everyday lives. Garbology is what we call a branch of archaeology that looks at people *today* rather than in the past. The Garbage Project studies garbology. The Project's scientists have collected evidence about many human behaviors, from what we buy and eat to how we spend our leisure time.

The Garbage Project started at the University of Arizona in the early 1970s. It began from some projects conducted in a college anthropology class. The students were trying to see if there were links between various objects and how people behave. Some students chose to study garbage. The study's results were so interesting that the teacher of the class decided that investigating garbage could give us evidence about our own behavior.

*Materials Through Time*

# 30

Studies of our garbage often provide evidence that does not agree with what people say they believe or do. For example, in a national survey conducted by the U.S. Department of Agriculture, people said they are eating less junk food and more healthy foods. But data collected by The Garbage Project shows something different. By analyzing packaging and other food wastes from fresh garbage and landfills all over the U.S., the group found that people eat more junk food, such as sugar, candy, potato chips, bacon, and ice cream than they say they do! The Garbage Project also found that people eat less healthy food, such as cottage cheese, liver, tuna, and vegetable soup than they say they do.

Garbage studies like these cannot tell us whether people are fooling themselves or just want to present a positive image to others. In any case, The Garbage Project studies can help us to learn more about nutrition, food waste, and consumer habits than we could from just asking people about these subjects in a survey.

Garbage studies can also tell us about what people buy, read, have as hobbies, and even what people do for fun! As you can imagine, not everyone wants someone looking through their garbage. For this reason, all data collected are confidential. For over 20 years, The Garbage Project studies have challenged some beliefs about human behavior and the overall makeup of our garbage. This kind of information provides a foundation of evidence to help us make good decisions about waste problems and waste management.

*For further reading:* Rubbish! *by William Rathje and Cullen Murphy, New York: Harper Collins, 1992*

**Questions**

1. What people say and what they do, don't always agree. In your Journal give an example from The Garbage Project that supports this statement.

2. Why do you think The Garbage Project got these results?

3. Think of a question you could answer by investigating garbage.

*Activity 30*   13

# Materials Through Time

## Introduction

### What's in the Landfill?

The chart below compares what the public perceived was the greatest problem in garbage to the actual volume in landfills found by The Garbage Project.

## Challenge

Your challenge is to compare what the public reported was the greatest problem in garbage to the evidence unearthed by The Garbage Project. How would you explain the differences?

*Public Perceptions versus Evidence of Landfill Volume*

| Item | Results of 1990 Roper Poll of the Public | Actual Volume in Landfills as Found by The Garbage Project |
|---|---|---|
| Disposable diapers | 41% | <2% |
| Plastic bottles | 29% | <1% |
| Large appliances | 24% | <2% |
| Newspapers | 11% | 13% |
| All paper | 6% | <40% |
| Food and yard waste | 3% | 7% |
| Construction debris | 0% | 12% |

# Materials Through Time

### Hands-on Career Training

Many people follow interesting career pathways. For example, Sheli Smith is the curator of the Los Angeles Maritime Museum. Before that, she worked as an archaeologist in charge of the excavation of a ship that was buried in a part of Manhattan built as construction fill. The Garbage Project played an important role in her training for both these jobs. In the mid-1970s she worked for The Garbage Project, once telling a *Wall Street Journal* reporter that she sorted garbage "to relax." When she worked as a garbage sorter, Smith appeared on the television show "To Tell the Truth." In this television show, which was very popular in the 1960s and '70s, celebrity panelists tried to guess the profession of guests. Before the show, Sheli Smith had a special manicure. None of the panelists were able to discover her job. One of the panelists said that "No one with nails like that would ever sort garbage."

*Reference:* Rubbish! *by William Rathje and Cullen Murphy, New York: Harper Collins, 1992*

## Communication Through Time

Mayan inscription in stone, 514 A.D.

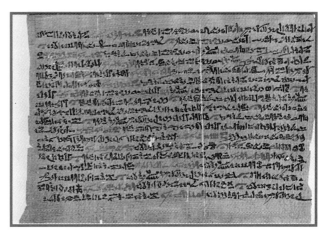

Medical papyrus, ancient Egypt

Printed material, 1440 to present

Computer diskettes, 1980s to present

# Activity 31: Properties of Materials

## Introduction

### Materials and Dwellings

Observe the pictures of buildings on the following pages very carefully. Most of them are dwellings—homes. Try to identify the materials used to make each building.

## Challenge

After trying to identify the materials used in each building, decide on the advantages and disadvantages of each material.

From the earliest of times, buildings have been made from materials that were easily available. The nature of the material used helped determine how strong the structure could be. It also affected how comfortable the people were inside. The structures we live in are meant to shelter and protect us from the changes in our environment—heat, cold, wind, rain, and snow.

# 31
## Properties of Materials

### Question

Look at each structure. How effective would each be in protecting its inhabitants? Relate your answers to what material(s) each is made of.

Buffalo hide

Wood

Metal, concrete, glass

## Properties of Materials

# 31

Brick

Stone, brick, tile

Plant material (reeds)

Stucco

Activity 31    19

# 31
## Properties of Materials

### Introduction
**Investigating Properties**

You will investigate a wide variety of materials. How can you tell the difference between one material and another? Properties of materials tell us about their similarities and differences. How do these properties determine how the materials are used?

### Challenge

Think about how the properties of different materials help determine the ways in which they can be used.

### Materials

*For each group of four students:*
- One stirring stick
- One 9-ounce plastic cup
- One glass slide
- One battery harness and bulb
- One 9-volt battery
- Samples of each of the following:
  - Aluminum
  - Carbon
  - Copper
  - Glass
  - Granite
  - Iron
  - Limestone
  - Melamine plastic
  - Polystyrene plastic
  - Tile
  - Wood

# Properties of Materials

## Procedure

Observe and record the following properties for each of the materials provided: color, luster, light transmission, texture, flexibility, resiliency, hardness, electrical conductivity, and density. Use the information below to help you with the tests for each property. Record your observations in a data table in your Journal. The more complex tests are described below.

### Safety

If a material is not easily flexed when you try to bend it, do not exert extra pressure to break or tear it. Be careful of sharp edges!
Wear safety eyewear.

### Hardness

To test the hardness of each material, determine if it is harder or softer than glass. Gently press the material across the surface of the glass plate. If a scratch appears that is not easily rubbed away, the material is harder than glass. If no scratch is seen, or if the scratch is easily rubbed away, the material is softer than glass.

### Electrical Conductivity

Test the electrical conductivity of each material by attaching the bulb and battery harness assembly to separate ends of the object. If the bulb lights, the material conducts electricity.

### Data Processing

Based on the properties of the materials, put them into groups. Each group must have one, two, or more properties in common. Record your groupings in your Journal.

### Density

You will compare the density of each object to the density of water. Fill the plastic cup half full of water and place the object in the cup. Check to see whether it sinks or floats. Push any material that floats under water with your stirring stick and see if it returns to the surface.

If it floats, it is less dense than water; if it sinks, it is more dense than water. As soon as you have tested the material, quickly remove it and dry it.

*Activity 31*

# Properties of Materials

### Introduction

## Olympic Materials

New materials have improved the way we dress, get to work, cook, and eat. In this reading you will discover how changes in materials have changed the performance of athletes in the pole vault.

### Challenge

After finishing this reading, think of other examples of new materials that have contributed to improvements in human performance.

In 1935, Brutus Hamilton, a track and field coach at the University of California at Berkeley, made some startling predictions. He said that under perfect conditions a man might someday run a mile in just over 4 minutes, high jump almost 7 feet, and pole vault over 15 feet—but those would be the limits. People laughed at his predictions because they seemed so impossible.

Well, we now know that it is possible to run a mile in well under 4 minutes. We also know that Coach Hamilton's predictions in the high jump and pole vault have been passed. The following charts show the Olympic records over the years. The improvement is highlighted for easy reference.

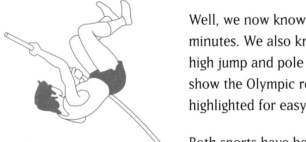

Both sports have benefited from improved training practices used by athletes. Synthetic track surfaces and lighter nylon running shoes have also helped. In addition, pole vaulters now use poles made of new composite materials including Kevlar and boron. The poles used today are very strong and flexible, a vast improvement over the rigid metal poles

*Properties of Materials*

# 31

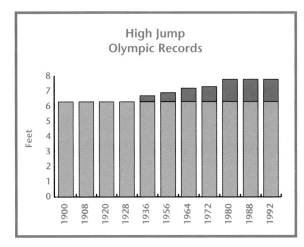

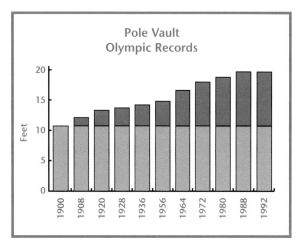

used in the early 1900s, the bamboo poles used in the 1920s and 1930s, and the plain fiberglass poles used until recently. (The athletes also must appreciate the soft foam pads they land on instead of the old pits full of sawdust used in the past.)

Pole vaulting today has changed from the early days of the sport. Computer models now project the maximum height possible given the speed of the runner at the instant when the pole is "planted" for takeoff, the kind of grip used, the twist of the vaulter as the pole flexes back from its bent position, and other variables. Unlike other aspects of track and field, all poles are not required to be the same. Different materials are favored by different athletes. Some athletes even treat the pole differently to improve their grip—some athletes use sticky tape or spray adhesives, and others smear a glue-like substance on the end of the pole. Athletes are constantly on the lookout for ways to use materials to improve their performance.

 **Question**

Examine the Pole Vault Olympic Records and High Jump Olympic Records graphs closely. Which graph shows the greater increase? Can you think of any reason for this difference?

Activity 31   23

# ACTIVITY 32: Conductors and Insulators

## Introduction

### Conductivity of Common Materials

Another property that affects how we use a material is its ability to conduct heat. The rate at which heat energy is transferred by a material is called its *thermal conductivity*.

## Challenge

Using the chart on the opposite page, compare the thermal conductivity of different materials. What properties do good thermal conductors have in common?

## Questions

1. Which two materials are the poorest conductors of heat? The best?

2. What can you generally say about the relationship between how much heat a material conducts and its density?

3. If all other factors are held constant, which one of the three little pigs would be the coolest in his house on a hot, calm summer day—the one that made his house of straw, sticks, or bricks?

## Conductors and Insulators

*Conductivity of Common Materials*

|  | Thermal Conductivity* | Density (g/cm³) |
|---|---|---|
| Aluminum | 164.0 | 2.7 |
| Common brick | 4.2 | 1.75 |
| Concrete | 2.5 | 1.4 |
| Copper | 278.0 | 8.92 |
| Fiberboard | 0.35 | 0.24 |
| Fiberglass, blanket | 0.20 | 0.05 |
| Glass, window | 5.5 | 2.5 |
| Gypsum, board | 1.0 | 0.8 |
| Ice | 12.5 | 0.9 |
| Iron | 55.0 | 7.86 |
| Limestone | 11.0 | 2.5 |
| Marble | 15.0 | 2.6 |
| Paper | 0.7 | 0.9 |
| Plaster | 4.2 | 1.8 |
| Plastics, foamed | 0.2 | 0.2 |
| Plastics, solid | 1.1 | 1.2 |
| Straw | 0.2 | 0.3 |
| Wood, balsa | 0.3 | 0.16 |
| Wood, oak | 1.0 | 0.7 |
| Wood, pine | 0.7 | 0.5 |

*Btu/(hr.)(ft.²)(°F/ft.)

# Conductors and Insulators

## Introduction

### Too Small, Too Red, Too Hot

This reading describes Thomas Edison's efforts to perfect the electric light bulb. His persistence is illustrated by the many materials he tried and the amount of time he spent trying to find the right design.

## Challenge

Your challenge is to identify both the variables that Thomas Edison kept constant and the one he changed as he tried to build the better light bulb.

"Too small, too red, too hot," said Thomas Edison in frustration one day in his lab over 100 years ago. Day after day, for over two years, he and his assistants experimented with thousands of different kinds of filaments (the part of a light bulb that actually gives off the light), trying to find just the right design. They were either so thin that they broke, or they gave off light that was too red, or they heated up so much that they melted. Even though the electric light was not his invention, he had a vision of a world lit at night by cheap, long-lasting electric lights. Edison even said that the day would come when only the rich would be able to afford candles. The problem was finding just the right materials that would do the job. His early failure led to success in the end.

## Conductors and Insulators

The basic idea was to send enough electricity through a conducting material to make it heat up and glow. If the material was shaped into a very thin, long strand, more heat and light would be given off. Edison knew that metals were good conductors of electricity, but he also knew that most would melt long before they gave off much light. He also wanted to produce a bright white light, not a dim red glow. The metal platinum held promise, though, so Edison announced that not only had he perfected the electric light but also that he had found an almost limitless source of platinum—a rare and expensive metal. He said that there were great amounts of platinum in the black sand of the Pacific Coast beaches. Neither claim turned out to be true. The platinum lights made at his workshop in New Jersey were very costly and burned out in just a few hours.

For many months he worked on designs that were sensitive to temperature. When the filament came close to melting, the electricity would automatically shut off. They were complicated, expensive, and never worked well. Many of the lights would often melt, and sometimes they would even explode, sending out a shower of sparks and glass. Those that did not explode would constantly flicker off and on. Certainly these were not going to replace the candle or gas lamp!

Edison then came up with the idea of using carbon (the black soot that he noticed on the inside of his gas lamp) for the filament of the light. Like metal, it conducts electricity, although not as easily. But the main advantage was that it didn't melt at high temperatures. He collected some of the carbon residue from a gas lamp and coated a thin cotton string with it. He then baked the coated string in a hot oven to harden it. To his dismay, it was extremely brittle and would break

## Conductors and Insulators

even before he could put it in a glass bulb. He had to put the carbonized string in a glass bulb, remove most of the air to form a vacuum, and then seal it shut so that it would work well.

For the next several years, Edison searched for the perfect material to coat with carbon—one that could be made into very thin strands but was not as brittle as the cotton string when it was hardened. He and his assistants tried cork, cardboard, wood shavings, coconut hair, leather, pasta, onion rinds, and even their own hair. Each experimental filament would take hours to make. It was not uncommon for Edison and his assistants to work 12 to 15 hours a day, sometimes even sleeping at the lab so that they could begin early the next morning.

In the end, he chose a particular type of bamboo for the light bulb filament, having tested over 6,000 different kinds of materials! The idea for using bamboo came to him when he picked up a Japanese fan sitting on his desk. Today we take the light bulb for granted, but its development is one of the great stories in the history of materials science.

Today very thin tungsten wire is used for the filament. It was chosen because it is easy to form into a tight coil, increasing the light output, and it has a high melting point. It can reach 4,500°F (2,500°C) and not melt.

### Questions

1. Edison once said invention is 99% perspiration and 1% inspiration. What did he mean?

2. How important was finding the right material for the filament of the electric light bulb? Explain your answer.

# Activity 33
## Preventing Corrosion

### Introduction

**Investigating Corrosion**

How well materials will last in the environment where they will be used is another important property of materials. For example, metals sometimes corrode, which affects their usefulness. In this activity you will investigate the effect of a number of factors on the corrosion of iron and other metals. You will consider the environmental tradeoffs involved in various approaches to the prevention of corrosion.

### Challenge

Your challenge is to determine the advantages and disadvantages, including environmental effects, of different approaches to preventing the corrosion of iron.

The rear fender of this car is corroded. What do you think happened to make the corrosion possible?

# 33 Preventing Corrosion

## Materials

*Materials for Group A:*
- One SEPUP tray
- Six small plastic cups
- One aluminum nail
- One stainless steel nail
- Two galvanized nails
- One iron nail
- One 30-mL dropping bottle of 50,000 ppm copper chloride solution
- Water
- Tweezers

*Materials for Group B:*
- One SEPUP tray
- Six small plastic cups
- One aluminum nail
- One stainless steel nail
- Two galvanized nails
- One iron nail
- One 30-mL dropping bottle of 50,000 ppm copper chloride solution
- Water
- One packet of salt
- One graduated cylinder
- One 9-ounce clear plastic cup
- One stirring stick
- Tweezers

# Preventing Corrosion

*Materials for Group C:*
- One SEPUP tray
- Four small plastic cups
- Two small pieces of steel wool
- One stirring stick
- Two iron nails
- Cooking oil
- Water
- One 60-mL flip-top bottle of denatured ethyl alcohol
- Tweezers
- Paper towels

*Materials for Group D:*
- Five 9-ounce clear plastic cups
- Four iron nails
- One 4"–5" piece of copper wire
- One 4"–5" strip of magnesium ribbon
- One 4" x 1/2" piece of aluminum foil
- Water
- Tweezers

# Preventing Corrosion

## Procedure

In this activity there are four different investigations identified as A, B, C, and D. Your teacher will assign you to one of the investigations.

A. *Testing Corrosion of Different Kinds of Nails in Water*

1. Use the SEPUP tray and five small plastic cups. Insert one plastic cup into each of the large cups in the tray, labeled A–E.

2. Place the aluminum, stainless steel, and iron nails in Cups A, B, and C, respectively. Place one galvanized nail in Cup D and one in Cup E.

3. Add enough water to fill the plastic cups in Cups A–D about three-quarters full.

4. Add enough copper chloride solution to fill the plastic cup in Cup E about three-quarters full.

5. Wait 15 minutes, and then record your observations.

6. A galvanized nail is an iron nail that is coated with zinc. What does the copper chloride solution appear to do to the nail? Record your ideas about what has happened to the galvanized nail in your Journal.

7. Now remove the plastic cup and the nail from large Cup E. Use the tweezers to hold the nail. Rinse the nail with water and put it in a clean plastic cup three-quarters full of water. Put this cup in large Cup E of your SEPUP tray.

8. Return the used copper chloride solution to the large bottle provided by your teacher.

9. Wait 15 minutes, and then record your observations in a data table in your Journal.

10. Follow your teacher's instructions for storing the samples. You will observe your results and present them to the class in the next class session.

### Safety

Wear protective eyewear. Some people may have an allergic reaction to the copper chloride solution, causing itching and redness to the affected area for a short time. Wash any exposed area with water for 2–3 minutes.

# Preventing Corrosion

### B. Testing Corrosion of Different Kinds of Nails in Salt Water

1. Prepare a solution of salt water by adding one packet of salt to a 9-ounce plastic cup. Use the graduated cylinder to add 60-mL of water to the salt. Stir thoroughly.

2. Use the SEPUP tray and 5 small plastic cups. Place one plastic cup into each of the large cups of the tray, labeled A–E.

3. Place the aluminum, stainless steel, and iron nails in Cups A, B, and C, respectively. Place one galvanized nail in Cup D and one in Cup E.

4. Use the salt water you prepared in Step 1 to fill each of the cups in Cups A–D about three-quarters full. Put the rest of the salt water aside. You will need it for Step 8.

5. Add enough copper chloride solution to fill the plastic cup in Cup E about three-quarters full.

6. Wait 15 minutes, and then record your observations.

7. A galvanized nail is an iron nail that is coated with zinc. What does the copper chloride solution seem to do to the nail? In your Journal, record your ideas about what has happened to the galvanized nail.

8. Next, remove the plastic cup and nail from large Cup E. Use tweezers to hold the nail. Rinse the nail with water. Take a clean plastic tasting cup and fill it about three-quarters full of salt water. Place the nail in the cup. Put this cup in large Cup E of your SEPUP tray.

9. Return the used copper chloride solution to the large bottle provided by your teacher.

10. Wait 15 minutes, and then record your observations in a data table in your Journal.

11. Follow your teacher's instructions for storing the samples. You will observe your results and present them to the class in the next class session.

*Continued on next page* →

**Safety**

Wear protective eyewear. Some people may have an allergic reaction to the copper chloride solution, causing itching and redness to the affected area for a short time. Wash any exposed area with water for 2–3 minutes.

**Preventing Corrosion**

**Procedure (continued)**

C. **Testing the Effect of Oil Coating and Size of Surface Area**

1. Take two small pieces of steel wool. Place one in a small plastic cup and add enough denatured ethyl alcohol to just cover the steel wool. Allow it to sit for 2 minutes. Then, pour the alcohol back into the bottle and rinse the steel wool and plastic cup with water several times. Use tweezers to hold the steel wool. Place the plastic cup containing the steel wool into large Cup A of a SEPUP tray.

2. Place the other small piece of steel wool into another plastic cup and set the cup in large Cup B of the SEPUP tray.

3. Take the two iron nails. Coat one nail with some cooking oil. To do this, place several drops of cooking oil on a paper towel. Wipe the oil-soaked towel over the nail so that it is completely coated. Be sure to rub the oil all over the nail. It must be completely covered with cooking oil. Place the uncoated nail and the oil-coated nail in two separate plastic cups. Put one of these cups in large Cup C and the other in large Cup D of the tray.

4. Fill each of the plastic cups about three-quarters full of water.

5. Wait 15 minutes, and then record your observations in a data table in your Journal.

6. Follow your teacher's instructions for storing the samples. You will observe your results and present them to the class in the next class session.

**Safety**

Wear protective eyewear. Denatured ethyl alcohol is toxic and flammable.

Activity 33

# Preventing Corrosion

### D. Testing the Effect of Contact with Other Metals

1. Take the copper wire and wrap it around one of the nails, as shown in the diagram below. Remove the nail from the wire. Stretch the wire slightly, as shown in the second diagram. Then insert the nail back into the spiral formed by the copper wire. The stretched wire should fit quite tightly around the nail.

Safety

Wear protective eyewear.

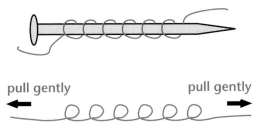

2. Repeat the same process. This time use the magnesium ribbon and a nail. Be extremely careful when you stretch the magnesium ribbon. It breaks very easily. Do not pull very hard on it.

3. Take the third nail and wrap it in aluminum foil. Roll the foil up tightly and then wrap it around the nail.

4. The fourth nail will not be wrapped in another metal. Take four 9-ounce plastic cups. Put each of the four nails into one of the cups. Next, fill each cup about one-third full of water.

5. Wait 10–20 minutes, and then observe the nails. Record your observations in a data table in your Journal.

6. Follow your teacher's instructions for storing the samples. You will observe your results and present them to the class in the next class session.

## Introduction

### Iron and Steel

Iron and steel are examples of widely-used materials. This reading describes how the development of the ability to use metals has changed the lives of people.

### Challenge

Your challenge is to think about the many ways that you use iron, steel and other metals in your life.

When horses were the primary means of long-distance transportation, horseshoes were one of the most important objects made of iron. Why would a horse need an iron shoe?

## Preventing Corrosion

People have been using metals for thousands of years. Gold and silver were used to make ornaments, but they were too soft to use to make weapons or tools. The Bronze Age began about 3500 B.C. with the discovery of methods for making bronze from copper and tin. But bronze didn't replace the bone and stone tools and weapons used by most people. Copper was too rare and costly. It wasn't until the discovery of methods for producing iron that a metal became widely available as a material. Iron, a strong, inexpensive, common metal led to major changes in the everyday lives of people living in what came to be known as Iron Age cultures. Iron is the fourth most abundant element in the Earth's crust and is found on every continent. This widespread availability of inexpensive iron has resulted in its position as the most widely used of all the metals. Today products made from iron and steel (an iron alloy) are used to produce almost everything we use.

People first began to use iron from meteorites in the period from 6000 to 4000 B.C. They chipped and hammered the meteorites to make weapons, tools, and ornaments. As people learned to harness fire by using bellows to push air containing oxygen into it, the temperatures they could obtain rose higher—high enough to extract iron from rusty-looking ores. In China, India, and Turkey, from 2500 to 1400 B.C., iron was produced by smelting ores. When the Hittite Empire in Turkey fell and the Hittite tribes dispersed, the knowledge of iron smelting from ore spread. About 1000 B.C. the Iron Age began. Iron smelting was independently discovered in Africa in approximately 500 B.C., near Lake Victoria (in an area that is now Burundi and Rwanda). Iron was used for tools, weapons, and eventually the armor of the Middle Ages. By A.D. 1000, iron was used to make tools for farming, such as horseshoes and plows. The increased production of tools and machines made possible by the use of iron led to the Industrial Revolution and great changes in farming as farms became more mechanized. In

# Preventing Corrosion

the 1860s, methods for producing large quantities of steel, an iron alloy, led to further revolution in the production of iron-containing products and permitted mass production by steel machinery and tools.

Iron is generally used in the form of steel—iron combined with other metals or carbon. The addition of these other elements allows the iron and steel industry to produce an endless variety of products to meet special needs. For example, the stainless steel used to make tableware is produced by adding chromium and carbon, in specific proportions, to iron. Imagine life without iron and steel products; automobiles, skyscrapers, household appliances, food cans, and the machines that produce the clothes you wear and the food you eat are just some of the products that would not be available!

## Questions

1. In ancient times, what was the first source of iron?
2. What are alloys? How are they used?
3. What differences would there be in your daily life if there were no iron or steel?

*Preventing Corrosion*

# 33

## Introduction

### Protecting Liberty

Now that you have performed your corrosion experiments, you will read about the efforts made to control the corrosion process on a very important national monument, the Statue of Liberty located in New York Harbor.

## Challenge

Your challenge is to apply your knowledge about the corrosion of metals to solve a practical problem.

She stands 151 feet tall, weighs 280 tons, and, on a clear day, is visible for 42 miles. Yet the Statue of Liberty is more than just a monumental figure. She has welcomed generations of immigrants and visitors to New York's harbor and has come to symbolize the United States of America to millions of people around the world. Protecting Liberty from the effects of corrosion, like protecting liberty itself, is not something we can turn our backs on.

When the French sculptor Frederic August Bartholdi designed the Statue of Liberty, he knew that he couldn't let the iron support structure inside the statue touch the copper skin on the outside. Why not? The skin is made of 300 thin copper plates joined together by some 300,000 copper rivets. The problem is *galvanic corrosion*. This occurs when two different metals—in this case, iron and copper—come in contact in the presence of an *electrolyte* (a solution that conducts electricity by the movement

# Preventing Corrosion

of ions). In galvanic corrosion, one of the two metals corrodes faster than the other. In this case, iron corrodes up to 1,000 times faster than copper!

To prevent galvanic corrosion, Bartholdi and his structural engineer Gustave Eiffel (who later designed the Eiffel Tower in Paris) put shellac and asbestos strips between the iron ribs and copper skin. This prevented the metals from touching so corrosion did not occur. But as Liberty's metal parts expanded and contracted with changing temperatures and the push of the wind, the insulator strips soon wore away, and the incompatible metals made contact. Seawater collected in the statue, and corrosion began quickly. Also, the rust building up between the iron and copper created a pressure great enough to pop out thousands of rivets, leaving holes in the skin for even more salt water to seep through.

After standing for almost 100 years in New York Harbor, the corrosion was so bad that the Statue of Liberty had to be closed for a massive rebuilding project. The iron bars holding up the copper skin had lost, in some cases, up to two-thirds of their original size.

As part of the rebuilding, all the flat iron bars in the skin-support system—more than 1,300 bars totaling 35,000 pounds—were replaced with stainless steel substitutes. Stainless steel is an alloy of iron, chromium, nickel, and molybdenum that is corrosion resistant. The old copper skin and new stainless steel support bars are now lined with Teflon—a slippery polymer that is an electrical insulator—to provide added protection against galvanic corrosion and to reduce problems caused when the statue expands and contracts in response to heat and cold.

*Adapted with permission from* ChemMatters, *April 1985, © 1985 American Chemical Society.*

*Preventing Corrosion*

**Questions**

1. What is galvanic corrosion? How does it affect the Statue of Liberty?

2. How can you prevent or slow down galvanic corrosion?

3. In some parts of the United States, roads are salted in the winter to melt ice. As you would expect, this increases the rate of corrosion of automobiles. Based on the investigations done in your class, what are some possible methods that could be used to prevent corrosion as a result of the salt? Which method would you recommend? Explain your choice.

# Activity 34
# Bag It! Paper or Plastic?

## Introduction

In this activity you will compare the properties of two materials commonly used in grocery bags—paper and plastic.

## Challenge

Design an ideal bag for your local supermarket. Identify what qualities (properties) your bag would have and why they are important.

*Bag It! Paper or Plastic?*

# 34

## Procedure

1. Work with your group of four students to list as many advantages and disadvantages as you can for the use of both paper and plastic materials for grocery bags. Make a table to organize your ideas.

2. Paper bags and plastic bags do not magically appear when needed and disappear when we finish using them. Where do they come from? Where do they end up? The entire manufacturing process, from the resources used to make a product to the final fate of the product, is called the *product life cycle.* In your Journal, copy the product life cycle diagrams below and write how you think having a diagram of the cycles helps you decide which bag to choose. What else do you want to know?

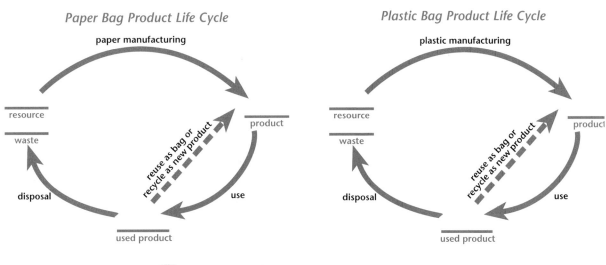

## Question

What do you think about the choices given in the cartoon—to "deplete oil reserves or cut down a forest?" Are these the only choices? Explain your answer.

Activity 34

# Materials in Clothing

Cotton dress from the Sudan, early 1900s

Apollo 17 spacesuit made of Teflon, mesh, and a neoprene-like substance

Leather vest

Wool sweater and silk scarf

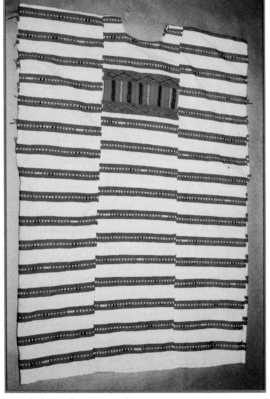
Wool dress from southwestern Mexico

Synthetic (plastic) rain slicker

# Activity 35 — Properties of Plastics

## Introduction

### Physical Properties of Plastics

You have examined the physical properties of a variety of materials. Now you will examine the physical properties of plastics.

## Challenge

Your challenge is to analyze the physical properties of four known plastics. Think about how these properties will affect how each plastic can be used.

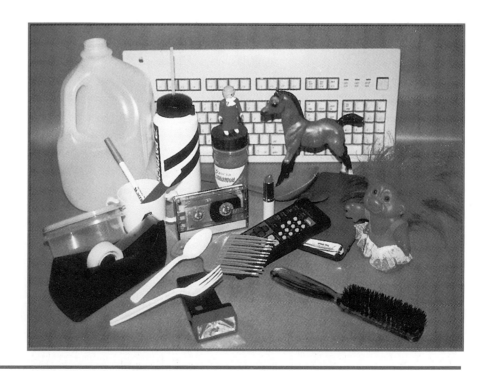

*Properties of Plastics*

# 35

## Materials

For each group of four students:

- Four plastic strips, one each, of:
    - polypropylene (PP)—blue
    - polyvinyl chloride (PVC)—green
    - high density polyethylene (HDPE)—red
    - polystyrene (PS)—yellow
- One paper clip, small nail, or penny

# Properties of Plastics

## Procedure

Each member of your group should test a different plastic strip by using the tests outlined below. Record your observations in a data table in your Journal like the one called "Testing Plastic Strips."

1. **Flexibility and Crease Color**

   Gently bend your strip back and forth to observe its flexibility (ability to bend) and the color of the crease that is produced. Record your observations.

2. **Hardness**

   The hardness of a material is its resistance to being scratched. Test for hardness by using the end of a paper clip, nail, or penny to see if it will scratch the plastic. Use only gentle pressure; do not gouge the sample! Record your observations.

3. **Acetone and Heat**

   Watch your teacher demonstrate the effects of acetone and heat on each of the four types of plastic. Record your observations.

*Testing Plastic Strips*

| | | Student Tests | | | Teacher Demonstration | |
|---|---|---|---|---|---|---|
| Plastic name | Color | Flexibility (excellent, good, fair, poor) | Crease Color | Hardness (scratches) yes or no | Effect of Acetone | Effect of Heat |
| Polypropylene (PP) | blue | | | | | |
| Polyvinyl chloride (PVC) | green | | | | | |
| High density polyethylene (HDPE) | red | | | | | |
| Polystyrene (PS) | yellow | | | | | |

*Properties of Plastics*

## Introduction
### Density

Density is a physical property of all materials. It is a comparison of the weights of the same volume of different materials.

## Challenge

Determine the relative densities of the four plastics by comparing whether they float or sink in various liquids.

## Materials

*For each group of four students:*

- One package salt
- Four plastic graduated vials, with caps, labeled 1, 2, 3 and 4
- Tap water
- One 9-ounce plastic cup
- One 60-mL bottle of denatured ethyl alcohol
- One dropper
- Four small, square plastic pieces, one each, of:
    - Polypropylene—blue
    - Polyvinyl chloride—green
    - High density polyethylene—red
    - Polystyrene—yellow
- Tweezers

Activity 35

# Properties of Plastics

## Procedure

 **Safety**

Denatured ethyl alcohol is toxic and flammable. Wear safety eyewear.

1. Label the four vials: 1, 2, 3, and 4.

2. Measure 10 mL of denatured ethyl alcohol into Vial 1. Place the cap on the vial.

3. Measure 5 mL of denatured ethyl alcohol and 5 mL of tap water into Vial 2. Place the cap on the vial. Shake well to mix.

4. Measure 10 mL of tap water into Vial 3. Place the cap on the vial.

5. Measure 10 mL of water into Vial 4. Pour in one package of salt. Place the cap on the vial and shake well to mix.

6. Each student in the group should test all four plastic squares in one of the four solutions. Prepare a data table to record your group's observations. The first student should place all four plastic squares in Vial 1, cap the vial, and shake gently to eliminate the effect of air bubbles or surface tension.

7. Everyone in the group observes the experiment and records their observations in a data table.

8. The person who tested the squares now removes them by using tweezers and placing them on a paper towel. Blot the squares with the paper towel and give them to another member of your group to test in Vial 2. Continue this until the plastics have been tested in all four solutions. Be sure your data table is complete.

# Properties of Plastics

### Data Processing

1. Which plastic was the least dense? What is your evidence?
2. Which plastic is the most dense? What is your evidence?
3. Place the four plastics on this scale of specific density.

*Density Scale (Water = 1)*

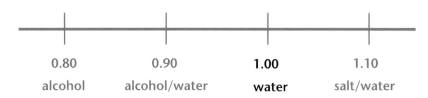

### Questions

1. Which plastics were the most flexible? Name a product in which this is a desirable quality and explain.
2. Which plastic was the least flexible? Name a product in which this is a desirable quality and explain.
3. You are a toy designer. Why would you want to know about the crease color of a plastic you might use in a toy?
4. Which plastic(s) scratched? How will this property affect their use?
5. Which plastic(s) were affected by acetone?
6. You are the design engineer for containers of nail polish remover (high in acetone). Of these four plastics, which one(s) would you consider using? Why?
7. Which plastics could be used for "dishwasher safe" food-storage containers? Explain.

# 35

**Properties of Plastics**

*A List of Common Polymer Products*

The following is a list of common items which contain polymers. In your Journal, list any of these items that you have encountered in the past seven days.

*garbage bags • insulated foam cups • milk jugs • toothpaste tubes • sandwich bags • fast-food containers • shoestrings • motor oil bottles • snack food bags • 2-liter soda bottles • plastic knives, forks, and spoons • shampoo bottles • shoestring tips • food wraps • disposable razors • egg cartons • thread spools • margarine tubs • disposable diapers • beverage boxes • coffee stirrers • insulation • caulking • shrink wrap • bubble pack • Silly String • Silly Putty • Slime • cellophane tape • epoxy glue • Super Glue • dishpans • plastic glasses and cups • pens • plastic dishes • Teflon cookware • food storage containers • combs • toothbrushes • ceiling light covers • tabletops • chair seats • carpets • refrigerators • telephones • floor tiles • synthetic fabrics for clothing • rubber soles • bicycle and automobile paint • tires • windshields • dashboards • floor mats • vinyl tops • bicycle handgrips • reflectors • shoe boxes • sweater boxes • vinyl wall coverings • sunglasses • glasses • compacts • contact lenses • pencil cases • lipstick tubes • Chap Stick tubes • hair sprays • windbreakers • raincoats • panty hose • galoshes • umbrellas • rubber gloves • sweaters • Nerf balls • Frisbees • snorkels • swim fins • wet suits • volleyballs and nets • basketballs • racquetballs • tennis balls • tennis racquets and strings • guitar strings • balloons • rubber bands • credit cards • portable radios • computers • watch faces • safety glasses • false teeth • hearing aids • lunch boxes • coffee mugs • Thermos bottles • lunch trays • flowerpots • supermarket meat trays • microwave cookware • lawn chairs • welcome mats • Astroturf • Velcro • football helmets • football pads • cleats • footballs • shuttlecocks • hockey pucks •*

Activity 35

## Properties of Plastics

*buttons • erasers • thread • wigs • false eyelashes • surfboards • parachutes • sailboats • sails • Pontiac Fiero • Corvette • Honda CRX • gears • Lucite sculptures • playing cards • floor waxes • furniture polishes • sousaphones • clarinets • flutes • recorders • records • tape recorders • videotapes • audiotapes • computer discs • luggage • typewriter cases • typewriter ribbons • metallic balloons • Habitrail • flea collars • index tabs • life rafts • model planes • model cars • jewelry • Colorforms • pacifiers • baby bottles • baby rattles • cushions • foam rubber pillows • pillowcases • exercise mats • photographic film • photographs • flashcubes • decorative fruit • plastic flowers • mannequins • street signs • store signs • book bags • chemistry classroom desks • school desks • knapsacks • rulers • protractors • overhead projectors • transparencies • movie film • slides • dust brushes • test tube brushes • rubber tubing • Tygon tubing • test tube racks • foam cups • fishing line • fishing bobbers • tackle box • waders • Weed Whacker • sneakers • vitamin capsules • sofas • safety glasses • sponges • seat cushions • particle board • hair dryers • boats • life preservers • hair curlers • Ping-Pong balls • tents • windshield wipers • shower doors • electrical tape • folding doors • bubble gum • house paint • ice chests • car batteries • toilet seats • water pipes • zippers • awnings • cameras • pianos • Band-Aids • mops • curtains • handles • tent pegs • plywood • picture frames • yarn • rope • shoe polish • fan belts • paint brushes • vinyl siding • checkers • flags • Rubber Duckies • puppets • earphones • whistles • patio screens • trash cans • attaché cases • wallets • motorcycle helmets • ashtrays • coasters • ice cube trays • hang gliders • sandals • skateboard wheels • microfilm • measuring tapes • extension cords • sun visors • venetian blinds • butane lighters • trophies*

*Adapted from* Polymer Chemistry, *1989; used with permission from the National Science Teachers Association, Arlington, VA.*

# 35 Properties of Plastics

## Introduction
### Identifying Unknown Polymers

In this activity you have learned about the physical properties of plastics. Now, you will use the data you have gathered to help you identify some unknown plastics.

## Challenge

Your challenge is to identify the unknown plastic sample the teacher provides.

## Materials

For each group of four students:
- One package salt
- Four plastic vials with caps, labeled 1, 2, 3, and 4
- Tap water
- One 60-mL bottle of denatured ethyl alcohol
- One dropper
- Plastic containers brought from home (optional)
- Tweezers

For each student:
- One strip of one of the four unknown plastics
- One small piece of one of the four unknown plastics

Activity 35

*Properties of Plastics*

## Procedure

Although you are sharing materials, you will work independently and identify your piece of plastic. Your group should prepare the vials for the density test as you did before.

### Testing the Unknown Plastic Sample

**Safety**
Denatured ethyl alcohol is toxic and flammable. Wear safety eyewear.

1. Using the strip and small square of plastic your teacher gives you, perform whatever tests are necessary to identify the plastic. It is one of the plastics you have studied, but it has a different color. Record the procedures you follow in your investigation and your observations on a separate piece of paper. Think about how you will organize the data you collect so that it is clear.

2. Ask your teacher to demonstrate the effects of acetone and heat on these plastics. Record your observations.

### Data Processing

1. Use what you have learned about the properties of the four plastics to help identify this unknown piece of plastic. Defend your decision by using your data and explain how you know your unknown is the plastic you have chosen, and not the other three.

2. Which of the four plastics from class that you have tested would be the best choice for the following applications? Give evidence from your investigations to support your decision.

    a. Thin film shrink-wrap material
    b. Hinges on a plastic ice chest
    c. Squeeze bottles
    d. Ocean buoys

*Continued on next page →*

*Activity 35*

## Properties of Plastics

**Optional Procedure** — Testing Plastics from Home

1. Identify a small sample of the plastic material you brought from home. Your teacher will help you cut up the sample into smaller pieces in order to more easily perform the tests. Record your observations in your Journal. Think about how you will organize the data you collect so that it is clear.

2. Explain your identification as you did for Data Processing Question 1 above.

# Activity 36

## Synthesizing Polymers

### Introduction | Cross-linking a Polymer

In the previous activities you investigated the properties of common plastics. The plastics you examined are a type of polymer, a substance whose molecules are made out of long chains of smaller molecules. Now you will become a chemical engineer and cross-link the polyvinyl acetate polymer found in white glue.

### Challenge

Your challenge is to investigate the properties of the cross-linked polymerized material you create from white glue and sodium borate and suggest possible uses for it.

### Materials

*For each group of four students:*
- One 120-mL bottle of white glue
- One 30-mL bottle of 4% sodium borate solution

*For each pair of students:*
- One 9-ounce clear plastic cup half filled with water
- One stirring stick
- Two graduated cups
- Paper towels

# 36 Synthesizing Polymers

## Procedure

**Safety**

Be sure to wear protective eyewear. Immediately clean up any spilled materials with water.

1. Set up your investigation report.

2. Pour 10 mL of white glue into one of the graduated cups. List the properties of the glue in your data table.

3. Pour 2.5 mL of the sodium borate solution into the other graduated cup. List the properties of the sodium borate solution in your data table.

4. While stirring with a stirring stick, add the sodium borate to the white glue. Closely observe what happens, including any evidence of a temperature change. Continue to stir until nothing further happens.

5. Place the polymerized material in the cup of water to wash off any excess white glue. Remove your polymer from the water and put it on a paper towel. Gently squeeze the material inside the paper towel to absorb the excess water. (Note: Avoid getting your polymer on the table, on the floor, or on your clothes).

6. Observe the properties of the new material. Make a list of them. Focus on the following categories:

   a. *Stickiness* (How does it stick to different kinds of surfaces?)

   b. *Bounciness* (How well does it bounce?)

   c. *Stretchiness* (How well does it pull apart when pulled both quickly and very slowly?)

7. Wash your hands and lab equipment thoroughly. Your teacher will instruct you on what to do with the polymer.

# Synthesizing Polymers

 **_Data Processing_**

1. How is the material you made different from the white glue?

2. The synthesis of a polymer involves a chemical change. What evidence do you have that a chemical change occurred?

3. What could you use this new material for?

4. What physical properties would you like to change in the material? Why?

# 36 Synthesizing Polymers

## Introduction | Designing a Material to Meet a Need

Plastics are rarely used in their pure form. Other materials are usually added to polymers in order to change their physical or chemical properties. In this way, plastics can be designed for specific uses. Adding materials in this way is called *compounding*.

## Challenge

Your challenge is to experiment with compounding the polymer you created previously to make it more suitable for a particular use.

## Materials

*For each group of four students:*
- One 120-mL bottle of white glue
- One 30-mL bottle of 4% sodium borate solution
- One 30-mL container of calcium carbonate powder
- One 60-mL bottle of liquid starch

*For each pair of students:*
- One plastic cup half filled with water
- One stirring stick
- Two graduated cups
- Paper towels

# Synthesizing Polymers

## Procedure

### Part One: Creating a Compounded Polymer

1. Work in pairs to do the investigation but set up your own investigation reports.

2. Add 5 cc (cc, or cubic centimeters, are used to measure powders) of calcium carbonate powder to one of the graduated cups. Then add 5 mL of liquid starch to the same cup and stir. You should now have a total of 10 mL of material in the cup.

3. Now, to this same cup, add 10 mL of white glue and stir. Observe the properties of the mixture. List them in your data table.

4. Pour 2.5 mL of the sodium borate solution into the other graduated container.

5. While stirring, add the sodium borate to the glue mixture. Stir until nothing further happens. Then rinse the new polymer, pat it dry, and observe its properties. List them in your data table.

### Data Processing

How is this compounded polymer different from the polymer you made from white glue alone?

*Continued on next page* →

## Synthesizing Polymers

**Procedure (continued)**

**Part Two: Investigating How the Ingredients (Reactants) Affect a Compounded Polymer**

1. After discussing your results within your group, try the same experiment again. However, this time leave out either the starch or the calcium carbonate powder.

2. Write up your experiment. Prepare to discuss the similarities and differences of this new compounded polymer with other student groups in class. If desired, let it stand overnight and observe any changes.

# Activity 37: Paper Clip Polymers

## Introduction — Making Models of Polymers

In the last activity you synthesized and compounded a polymer. Its properties were changed by what you added to it. This activity helps explain the behavior of the molecules in these substances. To help explain what we cannot see taking place, scientists use models. You will construct paper clip models to help you understand what took place as you combined ingredients and made the complex polymers. Keep in mind that a scientific model does not have to look like the real thing—it just has to act like it in an important way.

## Challenge

You will construct molecular models of polymer molecules that help explain some of the physical properties of your white glue polymers. Stick with it and work on!

## Materials

*For each group of four students:*

- Forty-eight paper clips (all the same size) in a 9-ounce clear plastic cup
- Six colored paper clips
- Two 9-ounce clear plastic cups
- One 60-mL wide-mouth bottle
- One plastic spoon

# Paper Clip Polymers

## Procedure

*Part One: Forming Polymers*

1. Put 24 unconnected monomer paper clips into the bottle. Slowly pour them from the bottle into one of your plastic cups. You may need to shake the bottle to help the clips move. Repeat this two or three times. Describe how quickly the clips come out of the bottle. Record your observations in your Journal in a data table like the one on the opposite page. Use the row labeled "Monomer" to record your observations.

2. Each member of your group should link 6 paper clips together to form a simple chain. Each clip represents an individual monomer molecule.

3. Link all four chains together to make a single chain of 24 paper clips. You have just made a polymer molecule! This chain represents the polyvinyl acetate (found in white glue) solution you began with in the last activity. (You would need thousands of paper clip monomers to make a realistic paper clip polymer. Such a model would be hundreds of times larger than the one you've just made.)

4. Now put your polymer in the bottle. Leave one clip hanging outside. Pour the chain into a plastic cup two or three times. Record your observations in the row labeled "Polymer."

5. Now put the polymer molecule into one of your plastic cups and the 24 separate monomer molecules into the other cup. Stir each with the spoon. Record your observations in the row labeled "Stirrability."

6. Remove the polymer chain from the cup. Pull on one end. Describe and record your observations in your data table.

Activity 37

## Paper Clip Polymers

*Comparing Model Monomer and Polymer Molecules*

|  | Pourability | Stirrability | Pulling |
|---|---|---|---|
| Monomer |  |  |  |
| Polymer |  |  |  |
| Cross-linked polymer |  |  |  |

### Part Two: Cross-linking Polymers

7. Separate the paper clip chain into four equal parts. Each part represents individual polymer molecules. Place the chains in four lines parallel to each other. Use two colored paper clips to make connections between the first and second chain. This is called cross-linking. In the last activity you cross-linked white glue to make a bouncy polymer. The sodium borate acted as a cross-linker, just as the colored paper clips do in your model polymer.

8. Repeat Step 7, using two colored clips between the second and the third chains and two between the third and fourth chains. Now you have a highly cross-linked polymer.

9. Carefully put your cross-linked polymer into the bottle, leaving one clip hanging out. Pour the polymer into a plastic cup. You may need to work at this. Record your observations in the column labeled "Pourability" for the cross-linked polymer.

10. Now try stirring your cross-linked polymer molecule in the plastic cup. Record your observations in the column labeled "Stirrability."

11. Pull on one of the paper clips in your model. Describe what happens to the rest of the paper clips.

12. Separate all of the paper clips and return them to the place designated by your teacher.

# Paper Clip Polymers

### Data Processing

1. If each paper clip represents a single monomer molecule, how many links or bonds must a molecule be able to make to form a chain like the one in Step 3 of your procedure?

2. Explain what causes the difference(s) in pourability.

3. Explain the reasons for any differences you observed in the stirrability of the 24 separate molecules (monomers) and the polymer chain.

4. Would you expect polymer molecules that are connected in long chains to have different physical properties from single monomer molecules? Why?

5. The scientific term that describes how easily a liquid flows is called *viscosity*. Highly viscous liquids are thick, while liquids of low viscosity are runny or thin. How would the viscosity of the monomer compare to the viscosity of the polymer?

6. Write a sentence to describe what happens to the properties of a chemical substance as more and more molecules of that substance are polymerized (forming longer chain molecules).

7. How is viscosity affected by cross-linking? Explain. (Recall what happened to the white glue when you added the sodium borate solution in the last activity.)

# Paper Clip Polymers

### Polymers

As the bowling ball slowly inched toward the pins, the crowd of spectators covered their eyes and ears. It was utterly quiet. Only the sound of the ball hitting the front pin could be heard. Then, a deafening explosion rocked the room. Once again, all the pins had been destroyed along with the ball. The spectators were elated.

This could have been a true story had not the earliest plastic materials been changed and other new plastics invented.

Plastics have been called "the first new materials to be made in 3,000 years," since the age of the discovery of metals. The earliest plastics were made from natural polymers in materials such as wood and cotton. In England around 1860, Alexander Parkes used cellulose fibers extracted from wood to make the first crude plastic. However, its quality was poor, and it took a contest in America to perfect the new plastic material.

John Wesley Hyatt was aware of the large reward being offered to replace the very costly ivory used to make billiard balls. In 1870, he developed a superior, easily molded product. He won the contest and patented the new product, which he called celluloid. It soon became the basis for the earliest movie films.

Celluloid was clear and flexible and could easily be molded but was very flammable. Projectionists in theaters kept a large bucket of sand nearby to smother potential flames. Sometimes the celluloid billiard balls would smoke and make small explosions as they collided. Even false teeth were made from celluloid. One unlucky smoker needed a fire extinguished in his mouth! If no other materials had been found to replace celluloid, the bowling ball story, instead of being fiction, might be fact. Today, Ping-Pong balls are the only product still made from celluloid.

With the discovery of large amounts of petroleum in the United States in 1859, plastics came to be made from crude oil. These plastic materials were synthesized from the smaller molecules created in the refining of crude oils. Hence, they are called synthetic plastics. The first, true synthetic plastic was made by the American chemist Leo Bakeland in 1907. Bakelite is still used today as an electrical insulating material.

# 37

*Paper Clip Polymers*

## Introduction

### The Big Stretch-Off (optional)

You have observed how the linking and cross-linking of molecules affects the properties of a substance. You will now investigate the property of plastic bags that allows them to stretch in different directions. After investigating this property of bags, the paper clip model from the last activity will be used to explain this behavior.

## Challenge

Determine whether the direction in which the lines run in a plastic bag affects how the bag stretches. Use the paper clip model to explain this ability to stretch.

# Paper Clip Polymers

**37**

Students hold squares of plastic bags up to the light to determine the direction of the lines in the plastic.

## Materials

*For each group of four students:*

- Two 12-cm x 12-cm pieces of plastic trash bag
- One pair of scissors
- One metric ruler

Activity 37

# 37

*Paper Clip Polymers*

## Procedure

1. You have two squares cut from a plastic trash bag. Hold them up to the light. Carefully observe! Do you see lines running parallel to each other? You should see lines running in one direction.

2. Next you will stretch a strip along the lines. Have one member of your group take one of the squares and cut it into 6 2-cm-wide strips. Cut the strips along the lines so that the lines run from one end of the strip to the other (see the diagram below). Give one strip to each member of your group. Save the other two strips for repeating this step, if necessary.

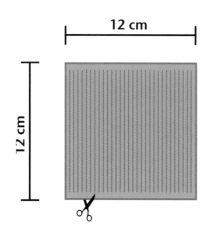
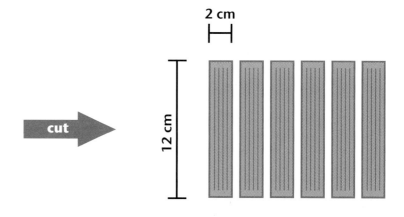

3. Grasp a strip in both hands so that each end extends over thumb and forefinger. Pull it gently. As you pull, look through the strip. What do you see happening? Use your ruler to measure how much each strip stretches before breaking. Record your observations.

*Paper Clip Polymers*

# 37

4. Next you will stretch a strip against the lines. Have a member of your group cut the other square into six 2-cm-wide strips the other way so that you cut across the lines in the strip (see the diagram below). Give one strip to each member of your group. Save the other two strips for repeating this step, if necessary.

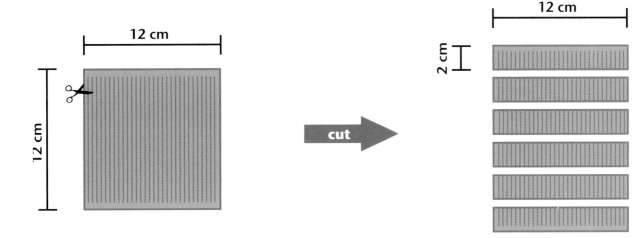

5. Grasp a strip in both hands so that each end extends over thumb and forefinger. Pull it gently. As you pull, look through the strip. What do you see happening? Use your ruler to measure how much each strip stretches before breaking. Describe how the plastic changed as it stretched before breaking. Record your observations.

# Paper Clip Polymers

 **Data Processing**

1. In your group, which strips stretched the farthest—the ones cut along the lines or the ones cut across the lines? Do the other groups agree with your group?

2. Describe any changes in appearance that your group observed as the strips were stretched along the lines and against them.

3. Describe how the polymer model could explain the stretching behavior of the plastic bag that you just investigated. It might help you to think about the polymer modeling we did in class with the paper clips and with our own bodies. You should try to come up with a hypothesis for how the arrangement of the polymer molecules in the bag could cause the stretching results you observed. Describe your model by drawing a picture and using words to explain it.

# Activity 38 Which Packing for Mike's Games?

### Introduction

**Popped Polymers**

Imagine that you are the head of the shipping department for a company called Mike's Games. Your company needs a packing material for its new video games. The suppliers of two packing materials have sent you information about and samples of their materials. The president of the company wants you to perform tests to compare the two materials and to make a recommendation to him about which polymer the company should use to pack the video games for shipping to customers.

### Challenge

Your challenge is to read the background information and advertisements on pages 75 and 77, and then test the two sample polymers to decide which material the company should use.

Some common packing materials

Activity 38   73

## Which Packing for Mike's Games?

### Information About Two Kinds of Foam

*Polystyrene Foam*

Polystyrene was first synthesized in Germany in the early 1930s. To make the foam product, small spherical beads of polystyrene are mixed with steam and pentane, a material found in small amounts in gasoline, to cause them to expand. The beads soften and allow the pentane to penetrate inside. The beads containing the pentane are treated with steam, and then are injected into a mold before a final application of steam is applied. By this time the spheres have expanded to nearly 27 times their original volume. The cooled product is extracted from the mold to become a cup or other foamed product. From 1 kilogram (2.2 pounds) of polystyrene beads, over 44,000 cups can be manufactured.

Formerly, substances called CFCs (chlorofluorocarbons) were used to expand the beads. These were discovered to destroy the Earth's protective ozone layer, which screens out ultraviolet light. The use of CFCs to expand foam products has been discontinued in the United States and is being phased out worldwide.

*Which Packing for Mike's Games?*

## Enviro-Pac

You can now pack all your products with our new environmentally friendly Enviro-Pac loose fill. It's made from a clean, nontoxic, and lightweight plastic (foamed polystyrene) that provides excellent cushioning for fragile items during shipment. It has greater crush strength than any other filler.

The new foam material contains no ozone-destroying chlorofluorocarbons (CFCs) that can damage the Earth's ozone layer. Instead, it contains a substitute (pentane) that the EPA views as environmentally safe.

It's superior to newspaper and other less resilient materials. It forms no harmful leachates in landfills because it is chemically inert—it doesn't react or breakdown. And it can be used in waste-to-energy incinerators (incinerators that use waste as fuel to produce energy) as fuel that contains more energy than coal.

Don't be misled by products that claim to be superior. For cost, ease of handling, moisture resistance, and customer satisfaction, choose Enviro-Pac for all your shipping needs.

## Which Packing for Mike's Games?

*Corn Starch Foam*

An alternate product made from a special kind of starch obtained from corn is called Eco-Foam. It is made from a type of corn that contains a different kind of starch from other corn. The starch is processed in a pressure cooker with a small amount of synthetic polymer and heated to form a gel-like material like gummy bears. When the pressure is released, the gel quickly foams to form the pieces that will be later used as packaging filler. When the foam is wet, the starch and binder quickly dissolve, and the material melts back into a gel.

*Which Packing for Mike's Games?*

## Nature Foam

Are you concerned about the effects of hazardous materials on the environment?

Now, there is an environmentally safe alternative to plastic foam. New Nature Foam uses material made from a renewable resource—corn—grown in America, which produces a loose fill material that contains no harmful products that will endanger the environment.

If you don't want to reuse the product, applicable disposal regulations may allow you to dispose of it in a compost pile, flush it down the toilet, wash it down the sink, or let the rain wash it away.

Why not use paper or popcorn for your packaging needs? Popcorn is messy, can turn rancid, has a higher weight per unit volume (it's more dense), and can attract rodents. Paper is heavy and requires extra time to pack into the box, increasing your costs of box packing and handling.

Switch now to the environmentally friendly Nature Foam. Although it does cost a little more per use than polystyrene foam, peace of mind makes it well worth it.

# Which Packing for Mike's Games?

## Procedure

1. Set up the investigation report form.

2. Read the background information and advertisements on pages 75 and 77. Look for evidence to help you make the decision. Record the evidence.

3. Decide which tests you and your partner will perform on the samples. You must do at least three tests.

4. Perform the tests on the polymer samples. Record the procedures you used and the results.

5. Check to see that you have done enough tests.

## Data Processing

Begin the investigation write-up. Your report should include a data table to present your results. Be sure to answer the following questions in the conclusion:

1. Which polymer did you choose for the shipping department of Mike's Games?

2. What are the reasons for your choice?

3. Are there any disadvantages to your choice?

4. What do you see as the tradeoffs in your decision?

5. What kind of evidence would help you make a better decision?

# Activity 39

## Comparing Garbage

### Introduction

Data about the volume and weight of solid wastes can provide useful information for making decisions about wastes. In this activity you will compare data about the garbage you generate with data about garbage collection nationally.

### Challenge

Compare your data with national data about garbage categories.

## Procedure

1. Use the Student Sheet to fill in the data about your personal garbage.

2. With your group, estimate the percentage of the total represented by each category. Record your group estimates on Student Sheet 39.

## Data Processing

1. Compare your class data with the data from the national landfill graphs. How are they alike? How do they differ?

2. How would you explain the differences?

3. If waste is placed in a landfill, what is more important—the volume or the weight?

4. What are some of the major concerns when making a decision about what to do with the total waste that is accumulated in a community?

## Comparing Garbage

*Composition of Municipal Solid Waste*

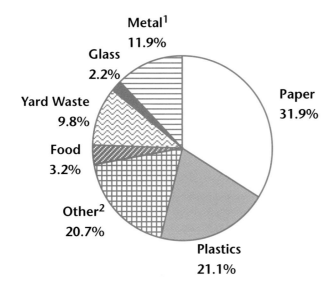

**Volume**

|  |  | V | W |
|---|---|---|---|
| [1]Metal: | Aluminum | 2.2 | 1.0 |
|  | Iron-like metals | 8.9 | 6.4 |
|  | Misc. | 0.8 | 0.0 |
| [2]Other | Rubber/Leather | 6.1 | 2.7 |
|  | Textiles | 6.4 | 3.3 |
|  | Wood | 6.8 | 7.3 |
|  | Misc. | 1.4 | 3.5 |

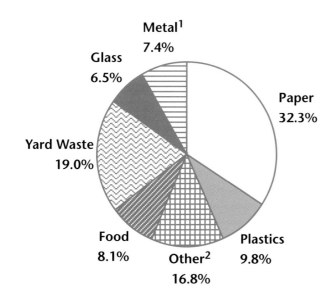

**Weight**

Source: Franklin Associates, Inc. 1992

# Activity 40: Investigating Sanitary Landfills

## Introduction

### Landfills

Most of our wastes end up in landfills. In the first two parts of this reading, you will learn more about the history of landfills and the construction of modern sanitary landfills. The third part discusses the possible environmental impacts of landfills.

## Challenge

Use the information in the reading, along with the investigations, to help you understand the advantages and disadvantages of sanitary landfills.

### Part One: An Historical Look at Landfills

Today, the average American discards 3.6 pounds (1.6 kg) of municipal solid waste each day, or nearly 1,300 pounds (590 kg) per year. This is approximately twice the amount that the average citizen of Germany discards. As a nation, we discard over 200 million tons (140 million metric tons) yearly. Every day in the United States 63,000 garbage trucks find their way to the nation's 5,200 landfills. If the trucks were all arranged in a single line, the line would stretch over 373 miles (approximately 600 km) bumper to bumper. If current practices continue, our annual solid wastes will total nearly 190,000,000 tons by the year 2000!

We live in a "throw-away" society. Whether we call it garbage, trash, refuse, or rubbish, it all boils down to the same thing: materials that someone no longer wants and wishes to see magically go away. As a society, we want all of the wastes simply to disappear, but where is "away"? We want the products created

*Investigating Sanitary Landfills*

by our technology, but we do not want the wastes created as part of their manufacture. This problem has resulted from the failure to connect the technological processes that we use to create products with the wastes that are created and have to be disposed of in some way. How did this begin?

Dumps and landfills have been the traditional answer to the problem of how to dispose of our wastes. In ancient Athens 2,500 years ago, it was decreed that wastes be transported at least one mile beyond the city gate. In colonial times, waste sites were located outside of villages or the wastes were burned as fuel. These sites were an improvement over earlier practices of dumping anywhere, but as the populations of cities increased and dumps grew larger during the Industrial Revolution, they soon created problems.

Some of the first attempts at constructing more modern landfills in the United States occurred in the Midwest in the early 1900s. Concern for public health was the main factor behind this change. By the 1930s, sanitary landfills were becoming more common. When the U.S. Army Corps of Engineers adopted sanitary landfills as the major disposal option for military facilities, many people were trained in their benefits and operation. Because of the odors and obvious pollution from the incinerators of the times, landfills came to be regarded as a highly desirable alternative. By 1960, about 1,400 American cities had sanitary landfills.

Today, modern landfills are carefully located, designed, and built to contain wastes and prevent their release to the environment.

### Question

How would you design a landfill that would address the pollution problems of the early dumps?

## Part Two: What Happens in a Landfill?

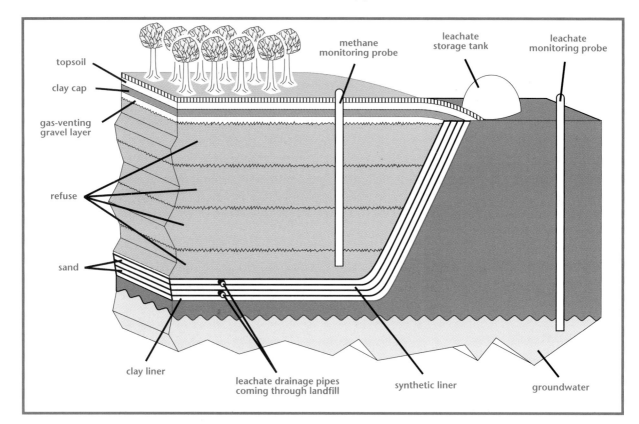

Studies of the geology and water sources of an area are now used to determine places where landfills can be located without causing contamination of ground and surface water. The landfills are usually lined with several feet of dense clay and then sealed with thick layers of plastic to prevent leaks. Each day, several inches of soil are spread over the landfill to prevent release of odors and to prevent rainwater from seeping in to form leaking materials called *leachates*.

As leachates collect in the bottom of the landfill, they are pumped out and then are collected and treated. The treated leachate is handled in a manner similar to sewage. The water is purified and the remaining solids (a muddy mixture called sludge) are dumped

*Investigating Sanitary Landfills*

in the ocean or a landfill, burned, or used as fertilizer. If the sludge is considered to be hazardous, it is sent to a hazardous waste disposal site.

Landfills also produce methane gas as a by-product of the breakdown of organic substances, such as lawn and food wastes, in the landfill. The newest landfills contain a system of pipes to collect the methane gas. This gas is sometimes vented to the air or burned. In other cases, it is purified and used as a fuel.

Although reducing the amount of waste dumped into landfills is one of the goals of modern waste management policies, landfills will always be with us. Even recycling produces wastes that must be landfilled. Landfills are controversial because of where they are located; no one wants to live near one—the so-called NIMBY (Not in My Back Yard) syndrome.

**Questions**

1. What is the purpose of spreading soil over the landfill each day?

2. How are the leachates that collect in the bottom of the landfill treated?

3. Name the various ways of disposing of the sludge remaining from the treatment of leachates.

4. What is produced as a by-product of the decay of organic wastes? How is it treated or disposed of?

5. What do you think should be considered when making decisions about where to locate landfills and how to manage the wastes? List the factors that should be considered and explain why each is important.

# Investigating Sanitary Landfills

*Part Three: Landfills and the Environment (optional)*

Leachates from landfills can have serious consequences. They can contaminate drinking water sources or have other harmful effects on the environment. The potential impact of leachates on the environment is illustrated by what can happen when landfills are built on or near wetlands. Wetlands are low-lying areas of land that are damp or wet most of the year. Marshes and swamps are examples of wetlands commonly found in the United States. Grasses, weeds, and other non-woody plants are common in marshes. In swampy areas, bushes and trees are also present.

Until the 1970s, wetland areas were often considered to be undesirable. By the early 1970s, more than 30% of the wetlands in the United States had been drained and/or filled. Large areas of cities, such as Boston, New Orleans, Miami, and St. Louis, are built on filled wetlands. Landfills were frequently located in wetland areas. The filled land could later be turned into desirable real estate. This was viewed as an added bonus of landfill construction.

Recently, our understanding of the ecological importance of wetlands has increased as a result of our greater knowledge of the impact of environmental changes. Wetlands act as a storage area for water, absorbing water during floods and acting as a reservoir of water in times of drought. In addition, wetlands serve as rich habitats for wildlife. Many species of animals, especially birds, obtain food and shelter and reproduce in wetland areas.

The leaching of toxic materials from landfills sited near wetlands, rivers, streams, and the ocean has contributed to the bad reputation of landfills in some areas. Once leachates enter the wetland areas (or other aquatic areas), they are able to spread into the environment. The problem can be magnified if the toxic substances are concentrated by living organisms in the food chain. If toxic substances are taken in by organisms at the base of

# Investigating Sanitary Landfills

the food chain, they can then become concentrated when eaten by organisms further up the chain. This is the case with some toxic heavy metals, like lead and mercury. Once they are taken in by algae and plankton, they cannot be broken down and excreted. They build up to higher concentrations in fish and invertebrates, like clams and crabs, that eat the smaller organisms. These are in turn eaten by birds and mammals, including humans. By now the levels of the heavy metals may have built up to harmful concentrations well above the threshold limits for humans and other organisms. The concentration of lead in fish can be 50 times greater than that in the water, and mercury can be concentrated up to 33,000 times!

Wetlands and other areas near surface water or aquifers are no longer considered appropriate places for landfills. Construction of landfills currently involves careful selection of a site as just one of many steps taken to prevent release of leachates into the environment.

## Questions

1. Until the 1970s what happened to many of the wetland areas in the U.S.?

2. Why are wetlands important?

3. What is the major problem from landfills located near wetlands, rivers, and the ocean?

4. What greater problems result when toxic substances leach into aquatic areas?

5. What happens to toxic heavy metals when they are eaten by aquatic organisms?

# 40 Investigating Sanitary Landfills

### Introduction
## Making a Model of a Landfill

Some materials, when "thrown away" carelessly, can leak into the surrounding soil and eventually get into the groundwater. These leaking materials are called leachates. A variety of precautions are taken to prevent materials from leaking out of landfills.

### Challenge

> Examine leachates from both a simulated unregulated dump and a regulated landfill in order to understand issues related to the safety of landfills.

### Materials

*For each group of four students:*

- Two SEPUP trays
- Two pieces of white paper
- Two paper towels
- Two droppers
- One 30-mL bottle of each of the following:
    - 20,000 ppm iron (III) nitrate solution
    - 0.1 M potassium thiocyanate solution
    - 0.5 M hydrochloric acid (HCl) solution
    - Water

# Investigating Sanitary Landfills

## Procedure

Your teacher will set up a demonstration of two leachate tubes to simulate an unregulated dump and a sanitary landfill. Iron (III) nitrate is used to simulate a toxic heavy metal. Student pairs will test the leachate samples prepared by your teacher. In order to determine the concentration of heavy metal in the leachates, you will need to set up a serial dilution of iron (III) nitrate to serve as a standard. You will determine the iron (III) nitrate concentration in the leachates by comparing them to the standards.

### Safety

Do not taste or touch the solutions. Wash any affected area with water for 2–3 minutes. Wear safety eyewear.

1. Prepare a serial dilution of the 20,000 ppm iron (III) nitrate in Cups 1–5 in a SEPUP tray. Start with 10 drops of 20,000 ppm iron (III) nitrate in Cup 1. When you have completed the dilution procedure, record the concentration and color of each dilution in Cups 1–5.

2. Obtain 10 drops of each leachate prepared by your teacher. Add these to Cups 6 and 7 of the SEPUP tray. Record the color of the leachates.

3. Test for iron by adding one drop of potassium thiocyanate to each of Cups 1–7. Record the color of the solution in each cup. Then add one drop of hydrochloric acid to each cup and record your observations.

4. Determine the concentration of iron (III) nitrate in the leachates by comparing them to the standards in Cups 1–5.

5. Set up a data table to organize and present your results. For each solution, you should record the color, both before and after testing, and the ppm of iron detected.

Activity 40

## Investigating Sanitary Landfills

### Data Processing

1. Observe the leachate tubes prepared by your teacher. What does the gravel/water layer represent?

2. What does the clay layer represent?

3. Explain what you can conclude about leachates from the simulated dump and landfill activity.

4. What risks, if any, are there in locating a dump on or near an aquifer used as a source of drinking or agricultural water?

5. What risks, if any, are there in locating a sanitary landfill on or near an aquifer used as a source of drinking or agricultural water?

6. Is it possible to operate a landfill that is completely risk free? Explain.

7. What materials would you dispose of in a sanitary landfill? Explain.

8. What personal action can you take to help make sure that wastes are disposed of properly—for example, that possibly hazardous substances are not put in sanitary landfills where they may endanger public safety?

# Activity 41
# Products of Hazardous Waste Incineration

## Introduction

First you will observe a simulation of the incineration of a hazardous waste that includes a toxic heavy metal. Then you will examine and test both the gaseous and solid products.

## Challenge

You will investigate the products of incineration. As you carry out the investigation, think about the advantages and disadvantages of hazardous incineration. Also think about what can be done to minimize the risks of operating a hazardous waste incinerator.

A hazardous waste incinerator

# 41

## Products of Hazardous Waste Incineration

### Procedure

*Part One: Up in Smoke*

1. Your teacher will burn a piece of paper containing simulated heavy metals and will collect the ashes. Observe the demonstration, record your observations, and answer Questions 1–3 on the next page in your Journal.

2. Your teacher will burn a second piece of paper containing simulated heavy metals and will collect the ashes and the smoke.

3. In your Journal record the results of your teacher's demonstration to test the smoke with bromothymol blue (BTB) indicator. Answer Question 4 and discuss your answer with your classmates.

4. In your Journal record the results of your teacher's demonstration to test the smoke with the potassium thiocyanate. Answer Question 5.

5. After viewing the videotape, "Incinerating Hazardous Wastes," answer Questions 6–8.

*Products of Hazardous Waste Incineration*

 ***Data Processing***

1. What are the two main products that come from the incineration of the heavy metal paper?

2. How effective was incineration in reducing the volume of the paper? What is your evidence?

3. How would you prove that the ash left behind had changed in weight? Do you think it weighs more, less, or the same as the unburned paper?

4. Why do you think the water that came from the bag used to collect the smoke behaved as it did with bromothymol blue (BTB) indicator?

5. Why did the water that came from the bag used to collect the smoke behave as it did with the potassium thiocyanate and hydrochloric acid?

6. What conclusions can you draw about the contents of the smoke based upon what you have seen in class?

7. What problems do the gases formed in incineration pose to the environment? What is a possible solution?

8. What problem does the ash formed in incineration pose to the environment? What is a possible solution?

# 41

## Products of Hazardous Waste Incineration

*Part Two: Investigating Residues*

### Materials

For each group of four students:

- Two pieces of filter paper
- One 30-mL dropping bottle of each of the following:
    - 0.1 M potassium thiocyanate solution
    - 0.5 M hydrochloric acid (HCl) solution
    - 20,000 ppm iron (III) nitrate solution
    - Water
- One ash sample
- One 1-cm x 1-cm piece of heavy metal paper
- One SEPUP tray
- One SEPUP filter funnel
- One stirring stick
- One dropper
- One paper towel
- One piece of white paper
- One small plastic spoon
- One pair of plastic tweezers

For each student:

- Journal records from the demonstrations

Activity 41

# Products of Hazardous Waste Incineration

## Procedure

Read through the entire experiment and prepare a data table to record your observations.

1. Fill large Cup A half full of water. This cup will be used for cleaning the stirring stick and dropper.

2. Place your SEPUP tray on a piece of white paper. Use the tweezers to put a 1-cm x 1-cm piece of unburned heavy metal paper in small Cup 1. Add 10 drops of water. Stir slowly for one minute.

### Safety

Follow your school policy retarding the use of safety eyewear. Do not taste or touch the solutions. Wash any affected area with water for 2–3 minutes.

3. Add one drop each of potassium thiocyanate and hydrochloric acid solutions to Cup 1. Stir the mixture and record your results in your data table. Clean your stirring stick.

4. Add a half spoon of ashes to small Cups 3 and 5. Break them into small pieces with your stirring stick. Add 20 drops of water to Cup 3. Stir.

5. Set up the SEPUP filter funnel over large Cups B and C. Fold two pieces of filter paper as directed by your teacher and place these in the funnels. Wet each thoroughly with 20 drops of water.

**Step 1**
Fold filter paper in half.

**Step 2**
Fold filter paper in half again

**Step 3**
Open with three thicknesses of paper on one side of the cone and one thickness on the other.

**Step 4**
Place paper in funnel and wet it with 20 drops of water to hold it in place.

*Continued on next page →*

## Products of Hazardous Waste Incineration

**Procedure (continued)**

6. With the dropper, remove as much liquid as possible from small Cup 3 and place it in the filter funnel over large Cup B. Clean the dropper. Allow the material to filter while you do the next part of the activity.

7. Add 20 drops of hydrochloric acid solution to small Cup 5. Stir. With the dropper, transfer all of the liquid in small Cup 5 to the filter funnel over Cup C. Allow the material to filter for 2 to 3 minutes.

8. Remove the SEPUP filter funnels to a paper towel. Discard the filter papers in the trash and clean and dry the filter funnels.

9. Record the color of the filtrates in large Cups B and C.

10. Add two drops of potassium thiocyanate solution to large Cups B and C. Add 1 drop of hydrochloric acid to large Cup B. Stir Cups B and C. Record the colors in your data table.

11. Use your data table from Activity 40 to find the approximate heavy metal concentration in each cup. Record this in your new data table.

12. Clean up as directed by your teacher and then answer the questions.

*Products of Hazardous Waste Incineration*

 **Data Processing**

1. How did the hydrochloric acid affect the heavy metal paper ash?

2. How does the concentration of heavy metal found in the ash and water filtrate compare to the concentration of heavy metal found in the ash and hydrochloric acid filtrate? How do they both compare to the unburned paper heavy metal concentration?

3. If incinerator ash contains toxic substances, such as heavy metals, it is commonly disposed of in special hazardous waste landfills. What are some problems these special landfills must solve in order to avoid the risk of contaminated groundwater? What can be done to solve these problems?

4. Batteries, some paints, and some pigment colors used in plastics contain heavy metals. What disposal policy would you adopt for your town if these materials were sent to the following sites: a community sanitary landfill? a hazardous waste incinerator?

**Extension**

Find out what methods, if any, are available in your area for the treatment and disposal of hazardous waste ash from heavy metals.

# Activity 42: Recycling Materials

### Introduction
**Investigating a New Plastic**

Recycling and reusing materials is an important component of the waste hierarchy and our goal to reduce the amount of waste we "throw away." This investigation looks at a new plastic which has some very unusual properties that enable it to be reused. It is used as a model for the recycling of materials generally.

### Challenge

To observe the new plastic's unusual properties, test it in an acid and a base. Based on your observations, determine whether or not there might be environmental benefits associated with the use of this type of plastic.

*Recycling Materials*

## Materials

*For each group of four students:*

One 30-mL dropping bottle of each of the following:
- 5% household ammonia solution
- 0.5 M hydrochloric acid (HCl) solution
- Water

*For each pair of students:*
- One SEPUP tray
- One piece of white paper
- One stirring stick
- One dropper
- Two 1-cm x 2-cm plastic film strips
- One paper towel
- One strip of pH paper

# 42

## Recycling Materials

### Procedure

Read through the entire procedure and make a data table to record your observations.

**Safety**

Do not taste or touch the solutions. Follow your school policy regarding the use of safety eyewear. Wash your hands after completing the activity.

1. Describe the plastic film (color, density, flexibility, the way light is reflected from it). Record your observations.

2. Place one of the pieces in large Cup A of your SEPUP tray. Add 50 drops of water. Stir with the flat end of the stick. Add a drop of the water to a piece of pH paper to check the pH of the liquid. Record your observations of the plastic film and the color of the pH strip.

3. Place the other piece of plastic in large Cup B of your SEPUP tray. Add 50 drops of household ammonia to large Cup B. Stir with the flat end of your stirring stick until no further change takes place.

4. Add a drop of the liquid in Cup B to a piece of pH paper to check the pH. Record your observations of the plastic-ammonia mixture and the color of the pH paper.

5. With the dropper, transfer the liquid contents of large Cup B to large Cup C.

6. While stirring, add 40 drops of hydrochloric acid solution to Cup C. Place a drop of the liquid in Cup C on a piece of pH paper to check the pH. If its color is green or blue, add 5–10 more drops of hydrochloric acid and retest the solution with another piece of pH paper. Repeat, as necessary, until the pH paper color is orange-to-red. You will know when you have been successful. You will see the mixture change from a milky white color to a lightly cloudy to a clear color. Record your observations.

7. Remove all of the reclaimed plastic, wash it and place it in large Cup D. Some of the plastic may stick to the stirring stick.

# Recycling Materials

8. Add 50 drops of water to Cup D. Touch a small piece of pH paper to the solution in Cup D. Record your observations.

9. Use the chart with the pH paper to estimate the pH of each of the solutions. Record your observations.

10. Remove the plastic from Cup D and place it between layers of the paper towel. Press down to squeeze out the water. Examine your recycled plastic. Try pressing it against a penny or shaping it with your hands. Record its appearance and properties in your data table.

11. Remove the plastic from Cup A to the paper towel. Compare the properties of this plastic to your reclaimed/recycled plastic. Record the differences in appearance and properties. Answer Questions 1–4.

12. Allow your recycled plastic to dry overnight and then re-examine its properties. Record its appearance and properties. Answer Questions 5–7.

## Data Processing

1. What effect does water alone have on the plastic? Why do we put the plastic in the water?

2. What happens to the plastic when ammonia is added?

3. What happens to the solution in Cup C when hydrochloric acid is added to it?

4. Write a sentence or two describing the relationship between pH and the properties of the plastic.

5. Compare the properties of the recycled plastic to those of the original. Did the recycling process affect the plastic in any way?

6. Could your recycled plastic be used again as a bag? What other uses might the recycled plastic have?

7. Do you think the solutions in Cups B, C, and D can be added directly to the city waste water (sewer) system of your community? Why or why not?

# Recycling Materials

## Introduction

### Plastics Recycling Today

Recycling plastics effectively requires some special approaches by the manufacturer, the consumer, and the recycler. Through watching and discussing a video and the sorting of plastic containers you bring into class, you will become more knowledgeable about plastics recycling.

## Challenge

After viewing the videotape, "Plastics Recycling Today: A Growing Resource," answer the following questions in your Journal. Think about what you want to discuss regarding plastics recycling.

## Questions

1. What are the advantages and disadvantages of making product containers from plastic compared to using other materials, such as glass or aluminum?
2. Write a sentence or two describing each phase of the plastics recycling process: collecting, sorting, processing, and remanufacturing the recycled plastic into useful products.
3. Is there a recycling center near your home that recycles plastic? Do you recycle plastics? Describe what you do to recycle plastics, or why you don't recycle them.
4. Is recycling plastics a good idea? Give reasons to support your position on this issue.
5. Using the codes and descriptions in the chart on the opposite page, sort the containers and/or records of containers you brought from home.

Activity 42

## Recycling Materials

*Plastics Resin Codes*

| Code | Name | Common Uses | Examples of Recycled Products |
|---|---|---|---|
| 1 | PETE—Polyethylene Terephthalate (PET) | soft drink bottles, peanut butter jars, salad dressing bottles | liquid soap bottles, fiberfill for winter coats, paint brushes, soft drink bottles, film, egg cartons, skis, carpets, boats |
| 2 | HDPE—High Density Polyethylene | milk, juice, and bleach bottles; grocery bags; toys; liquid detergent bottles | base cups of soft drink bottles, flowerpots, drain pipes, toys |
| 3 | V—Vinyl, Polyvinyl Chloride | shampoo bottles, clear food packaging, pipes | pipes, floor mats, hoses, mud flaps |
| 4 | LDPE—Low Density Polyethylene | plastic films, bread bags, frozen food bags, grocery bags | garbage can liners, grocery bags, multi-purpose bags |
| 5 | PP—Polypropylene | ketchup bottles, yogurt cups, margarine tubs, medicine bottles | paint buckets, ice scrapers, fast-food trays, automobile battery parts |
| 6 | PS—Polystyrene | videocassette cases, compact disc jackets, grocery store meat trays, coffee cups, prescription bottles, utensils | license plate holders, flowerpots, hanging files, food service trays, trash cans |
| 7 | Others, including mixed polymers | | plastic lumber |

Activity 42

# 42

## Recycling Materials

### Introduction

**Are We Recycling Enough?**

During World War II recycling was used extensively in this country. Today, there are actually fewer recycling (salvage) centers than in 1945. Is recycling in your community increasing? How can we help communities recycle more? This article may help answer some of these questions.

### Challenge

Recycling delays the day when a material's useful life finally ends. Why don't we do more of it? Think about ways to encourage people to do more recycling.

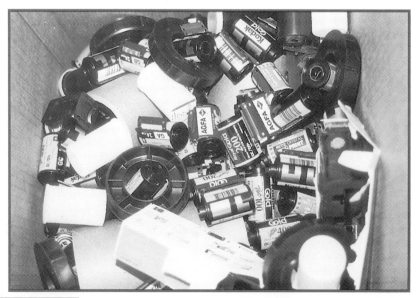

Film canisters can be turned in for recycling.

20,000 B.C. – bone, ivory sewing needles

Activity 42

## Recycling Materials

Recycling prevents useful materials from being burned in incinerators or buried in landfills. It can save energy and natural resources, provide useful materials from discards, and save landfill space.

Municipal solid waste can be thought of as a mixture of potentially valuable materials and fuels. Glass, plastics, paper, and metals are excellent candidates for recycling. Yard wastes can be composted and added back to the soil or incinerated for energy. Yet large amounts of recyclables are still ending up in incinerators and landfills, where they account for a large percentage of the total waste. For example, even though 50% of all aluminum cans are recycled, we still discard enough aluminum to rebuild our entire domestic air fleet each year. Why is this?

The answer depends on a number of factors, including the effort necessary to recycle, the costs of recycling, and sometimes the lack of a market for the recycled products.

Communities must examine the costs and benefits of recycling, considering questions such as: What markets exist for this material? How do the costs of recycling compare to the costs of incinerating or landfilling the wastes? For many cities, paying to recycle paper may be less expensive in the long run than landfilling. Local, state, and national governments are increasingly aware of this over-supply problem. Some have required the purchase of recycled paper. For example, in California, newspapers must contain at least 25% recycled newsprint.

Many cities and states are taking innovative approaches. Some states no longer allow yard wastes to be placed in landfills since many of these wastes can be composted. Some landfills are now being fitted with giant shredders to grind yard wastes into compost that can be distributed to citizens or sold as fuel. In many communities nationwide, citizens are charged directly for the amount of trash they generate. For example, they pay one rate for their first can. Additional cans cost extra. The result of this policy is an awareness of the economic costs of disposal. The policy has also helped to encourage residents to recycle, compost, and reuse. According to a study done by Duke University for EPA, it has helped reduce the amount of material sent to landfills by an average of 45%, helping these communities achieve the 25% reduction goal in landfilled wastes as set by the EPA for the entire country.

**Question**

Historically cars have been dumped in junk yards when their useful life is over. Based on everything you have learned about waste management in the last few activities, suggest a complete plan for the disposal of a car. Be sure to consider the car's metal body, plastic bumpers, lead-acid battery, interior, and other components.

# Activity 43

## Chemical Change: The Aluminum-Copper Chloride Reaction

### Introduction

Sometimes when two or more materials are mixed, they change to form new materials. You will explore chemical interaction by examining the reaction between aluminum and a solution of copper chloride. This is an example of a single replacement reaction that can be used to recover copper for reuse.

### Challenge

Your challenge is to find as much evidence of chemical interaction as you can in this reaction. How do the properties of the reactants and products of a chemical reaction compare to each other?

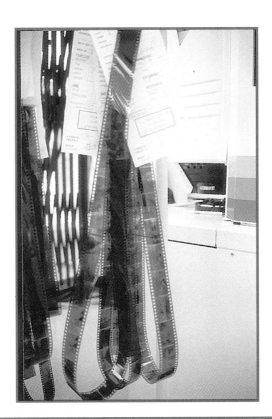

The silver from chemicals used to develop film is reclaimed and reused because it is too toxic (and too valuable) to dispose of in the sewer system.

# 43 Chemical Change: The Aluminum-Copper Chloride Reaction

## Materials

*For each group of four students:*

- One SEPUP tray
- One graduated container
- One 10-cm x 10-cm piece of aluminum foil
- Two 9-ounce clear plastic cups
- One plastic spoon
- One wooden or metal spoon
- One dropper
- Paper towels
- One 180-mL bottle of 50,000 ppm copper chloride solution
- One 30-mL dropping bottle of each of the following:
  - 50,000 ppm copper chloride solution
  - 5% household ammonia solution
  - Water

# Chemical Change: The Aluminum-Copper Chloride Reaction

## Procedure

1. Before you proceed, record some properties of the aluminum foil and the 50,000 ppm copper chloride solution in your Journal.

2. Carefully fold the aluminum foil in half, and then in half again. Place this in the bottom of one of the plastic cups.

3. Fill a second plastic cup 1/3 full of water. Place the plastic cup with the foil inside this cup.

**Safety**

Wear protective eyewear. Some people may have an allergic reaction to the copper chloride solution, causing itching and redness to the affected area for a short time. Wash any exposed area with water for 2–3 minutes.

4. Use the graduated container to measure 25 mL of copper chloride solution from the 180-mL bottle. Pour it onto the aluminum foil in the clear plastic cup.

5. Use the spoon to push the corners of the aluminum down so that the foil is completely covered. As the reaction proceeds, record as many observations as you can. Try not to include any interpretations.

6. While the reaction proceeds, set up a 1 to 10 serial dilution (as you did with the red food coloring in Activity 3) of the copper chloride solution. Start by putting 10 drops of 50,000 ppm copper chloride from the 30-mL dropping bottle in Cup 1 of a SEPUP tray. Continue the dilution through Cup 5. After the aluminum-copper chloride reaction seems to be finished, use a dropper to add 10 drops of the liquid from the plastic cup containing the aluminum reaction to Cup 7 of the SEPUP tray.

7. Add 5 drops of household ammonia to Cups 1–5 and Cup 7. What do you observe? Record your results in your Journal in a data table like the one on the next page.

*Continued on next page→*

Activity 43

# Chemical Change: The Aluminum-Copper Chloride Reaction

**Procedure  
(continued)**

Testing for Copper Concentrations

| Cup | Color of Solution ||  Concentration (in ppm) |
|---|---|---|---|
|  | After Water Dilution | After Adding Ammonia |  |
| 1 |  |  | 50,000 |
| 2 |  |  |  |
| 3 |  |  |  |
| 4 |  |  |  |
| 5 |  |  |  |
| 6 |  |  |  |
| 7 |  |  | approximately _____ |

8. Use your dropper to carefully remove the rest of the liquid from the plastic cup containing the aluminum reaction. Put it into a container provided by your teacher.

9. Using the plastic spoon, scoop up as much of the solid material from the clear plastic cup as possible onto a folded paper towel. Remove as much liquid as you can. Then, use the bottom of *a wooden or metal spoon* to press the solid material together. Observe and record what you see.

# Chemical Change: The Aluminum-Copper Chloride Reaction

### Data Processing

1. Copper compounds usually have a blue or green color. Turquoise is a compound of copper. How effective does the aluminum appear to be in removing the copper from the copper chloride solution? What is your evidence?

2. Describe the solid material. What do you think it is? Why? What is your evidence?

3. How would you go about proving or disproving that the gas generated in the aluminum-copper chloride reaction was water vapor (steam)?

4. What evidence did you observe that would help you prove a chemical reaction took place? Give three examples.

# 43 Chemical Change: The Aluminum-Copper Chloride Reaction

## Introduction

**It's Elementary and Shocking!**

What are the smallest things you can think of? In this reading you will discover how the idea of an atom was developed.

## Challenge

The idea of atoms began in the minds of philosophers from countries as far apart as India and Greece. Imagine the following conversation in Greece around 420 B.C.

"Hey Calicoes, have you heard the latest? A group of our fellow Greeks called Atomists say that everything we see is made of small particles called atoms! They say these atoms are so small that we cannot see them with our eyes. They claim that everything is made of empty space filled with atoms in constant motion, like tiny dancing particles of dust in a sunbeam! They believe that water and iron are different because they are made of different kinds of atoms. Could this be true, my friend? How can reason tell us about something we cannot see?"

The Atomists were a group of scholars who proposed that there were simple particles called atoms. They believed that what we call matter could be broken down into smaller and smaller bits until we arrived at a bit that was no longer divisible or able to be cut. They called this an atom—literally from the Greek word, *atome*, which means uncuttable. This bit-sized particle still retained all the physical and chemical properties of those large chunks of matter, such as a chunk of gold. They believed that the

# Chemical Change: The Aluminum-Copper Chloride Reaction

world was made of different kinds of atoms. By combining and recombining with each other, these different atoms produced material things and all the changes we see taking place in them. You may have never thought of this before, but when you ask such questions as: "Is this ring really gold?," "What's in my pizza?," "Why do things corrode?," and "Why do I change and grow older?," you are really asking the same question as the early Greeks, "What is the world made of and why does it change?"

For over 2,000 years the concept of atoms was not widely accepted. During those years, early chemistry developed from the work of alchemists, who believed that all metals were essentially the same. The alchemists devoted their lives to searching for ways to turn common metals such as lead into valuable silver and gold. Alchemists in Egypt, Greece, Turkey, and other parts of the Middle East laid the foundation for work that continued throughout Europe and Asia into the 18th century. Although these early scientists never did find a way to turn lead or iron into gold, some of them made important contributions to the study of science. For example, early Islamic scientists are credited with introducing careful record keeping and controlled experiments.

The ideas of the Atomists were revived in the 13th and 14th centuries when Islamic and Jewish scholars in Spain rediscovered and translated the books of the early Greeks. The theories of the Atomists and the ideas and work of other scientists laid the foundation for the modern atomic theory, developed by John Dalton in the early 1800s. Today, we believe that the individual particles that make up pizzas, stars, watches, whales, and water plus everything else, are composed of atoms. We call materials that are composed entirely of only one kind of atom *elements*. There are now 110 different elements known to exist. The last one (Element 110) was produced in a laboratory in Berkeley near the SEPUP headquarters in 1994. Only nuclear reactions can produce new elements or change one element into another. The alchemists

# Chemical Change: The Aluminum-Copper Chloride Reaction

were unable to turn lead into gold because lead and gold are different elements composed of different atoms. Although atoms cannot be changed into other atoms by normal chemical reactions, they can combine with each other to form molecules. Materials that have two or more different kinds of atoms are called *compounds*. There are so many compounds, and new ones being made all the time, that it would take many large books with tiny type just to list them!

Your body is made from many different combinations of a very few common elements. These are mainly hydrogen, oxygen, carbon, nitrogen, phosphorous, and sulfur. Other elements, such as calcium, magnesium, iron, sodium, potassium, and iodine, play an important role in our body's chemical energy plant, but are used in smaller amounts than those listed in the last sentence. Water is a compound made up of two hydrogen atoms and one oxygen atom. A single particle of the compound water is a molecule.

Now you can answer the ancient question that the Greeks posed over 2,400 years ago. The world is made of atoms that are in constant motion. It changes because these atoms combine with each other to make compounds. In turn, the atoms in compounds can break apart and recombine to form new compounds in a seemingly never-ending dance on the atomic level of change.

As scientists probed the atom more closely, they discovered that it was composed of even smaller particles that had electrical charges. Just as the poles of a magnet have north and south poles, atoms have positive and negative charges. The negatively-charged particles are called *electrons*. The positively-charged ones are called *protons*.

# Chemical Change: The Aluminum-Copper Chloride Reaction

It was determined that electrons are found near the outside of an atom, and the protons are found near its center. This would help explain an observation you have experienced many times. When objects are rubbed (like your shoes on the carpet on a dry day), they are able to transfer electrons easily because the electrons occur near the outside of atoms. These electrons jump, and you may get a shock or see a spark fly. This also happens when you pull a wool blanket away from a cotton sheet on a dry day.

So now we know atoms are made of smaller particles called electrons and protons. (A third particle that is found with protons in the center or nucleus of an atom is a *neutron*. It is neutral—neither positive nor negative.) Normally atoms (elements) have the same number of positive and negative charges; if they did not, we would receive a shock every time we touched anything! But elements, as we said before, are rarely found in their pure form in nature (gold, silver, platinum, copper, sulfur, and carbon are the only ones). Most elements combine with each other to form compounds in a chemical change.

One way this happens is by atoms losing or gaining electrons from each other. Common table salt is made up of sodium and chlorine atoms, but they have changed their elemental properties enormously. Sodium is a soft metal that explodes when immersed in water. Chlorine is a toxic, greenish-yellow gas. When these two elements come in contact with one another, each sodium atom loses an electron to a chlorine atom. See the diagram on the next page.

As you can see, sodium, which was neutral before, has one less electron. Before, it had the same number of electrons (- charges) and the same number of protons (+ charges). Now there is one less

# 43 Chemical Change: The Aluminum-Copper Chloride Reaction

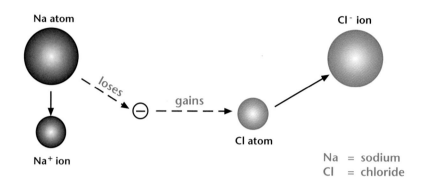

electron than protons. We say that the sodium atom has a net positive charge, and we no longer call it an atom but an ion. An ion is a charged atom (or group of atoms). We call this ion a sodium ion.

What happened to the chlorine? Remember, it has gained the electron that sodium lost. What will be the net charge on chlorine when it gains an extra electron? That's right! It will have a net negative charge, and we call it a chloride ion. The positively-charged sodium ion and the negatively-charged chloride ion now attract each other to form the compound NaCl.

This process of materials losing and gaining electrons takes place in chemical reactions constantly. In your experiment with aluminum foil and copper chloride, aluminum atoms lost electrons to become aluminum ions. Copper ions gained electrons to become copper atoms—the reddish-brown solid in your cup. Later, in the third part of our course, we will study how chemical reactions can generate a very useful form of energy called electricity.

# Chemical Change: The Aluminum-Copper Chloride Reaction

**Questions**

1. Explain how an element can become an ion.
2. What's the difference between an element and a compound?
3. Why does the material world change?

# Activity 44: Reducing and Recycling Hazardous Materials

## Introduction

In Activity 43, you found that aluminum metal removed copper from a toxic copper solution. Similar methods can be used to reclaim heavy metal wastes. Reclaiming copper serves a dual purpose—it reduces the toxicity of the waste stream and conserves a valuable metal. However, reclaiming the copper involves using another metal, which involves costs and the risks associated with disposal of the metal.

## Challenge

Your challenge is to investigate the effectiveness of three metals in reclaiming copper waste from a solution. Using the results of your investigation and additional information about the metals, you will choose a metal for reclaiming copper waste.

*Reducing and Recycling Hazardous Materials*

## Materials

*For each group of four students:*

- One 30-mL dropping bottle of 50,000 ppm copper chloride solution
- One 30-mL dropping bottle of household ammonia solution

*For each pair of students:*

- One SEPUP tray
- One dropper
- One aluminum washer
- One iron washer
- One piece of zinc shot
- One stirring stick
- One plastic spoon
- Paper towels

Activity 44

# Reducing and Recycling Hazardous Materials

## Procedure

1. Place the aluminum washer in Cup 1. Put the zinc shot in Cup 2. Put the iron washer in Cup 3.

2. Add 20 drops of 50,000 ppm copper chloride solution to each of the cups. Put 20 drops in Cup 4. This is the control.

3. Stir the contents of each cup every minute or so. Clean the stirring stick between each cup.

4. Observe what happens to each metal for 5–10 minutes. Record your observations in a data table you construct in your Journal. Be sure to include comparisons of the differences in results obtained with different metals.

### Safety

Wear protective eyewear. Copper chloride solution is toxic and corrosive. Avoid contact with skin and eyes. Wash any exposed area with water for 2–3 minutes. Some people may have an allergic reaction to the copper chloride solution, causing itching and redness to the affected area for a short time.

5. Using the plastic spoon, remove the pieces of metal from the cups and place them on a paper towel. Clean the spoon between each cup.

6. Make a record of your observations of the solutions left in each cup. Be sure to include comparisons of the differences in results.

7. Examine each of the metal pieces and describe what you observe. Be sure to include comparisons of the differences in results.

8. Using a dropper, transfer about 5 drops of each of the solutions to a clean cup in the tray. For example, transfer 5 drops from Cup 1 to Cup 5, from Cup 2 to Cup 6, etc. Clean the dropper between each cup.

9. Test for copper in the solutions by adding 2 drops of household ammonia solutions to Cups 5, 6, 7, and 8. Copper ions form a deep blue color or a blue-green precipitate in the presence of ammonia. Remember that Cup 8 contains the unreacted copper chloride solution (the control).

10. Record your observations. Your teacher will tell you how to dispose of each metal and clean out your tray. Complete the Data Processing section that follows on the next page.

# Reducing and Recycling Hazardous Materials

 **Data Processing**

Prepare a written report. Start your report on a clean sheet of paper. Include your name, the date, and a title for your report. Your report should have the following four components:

- A statement of the problem you were trying to solve.

- A description of the materials and procedure you used to solve the problem.

- A clear presentation of the results you obtained.

- An analysis of the results. Include your conclusions, any problems you may have had with the investigation, and any additional questions you would like to investigate related to the problem. Your analysis should include answers to the following two questions:

    1. Overall, which metal seemed to work best at removing the copper from solution? Describe completely your evidence for your answer.

    2. The table on the next page summarizes typical information about the costs and legal levels of iron, aluminum, copper, and zinc ions that may be disposed of in waste water. If you were to use one of these metals to remove the copper ions from the used solution, which one would you choose? Give reasons for your choice. Consider the results of your investigation as well as factors, such as cost, allowable disposal levels in the waste water, and possible uses of the metals for other purposes.

# Reducing and Recycling Hazardous Materials

*Information About Four Metals*

| Metal | Estimated Cost | Maximum Level in Waste Water | Availability |
|---|---|---|---|
| Aluminum | $0.90 per pound | not restricted | wide |
| Iron | $0.02 per pound | 100 ppm | wide |
| Zinc | $0.59 per pound | 5 ppm | wide |
| Copper | $1.33 per pound | 5 ppm | wide |

# Activity 45

## Source Reduction

### Introduction

**Put a Little Color Into Your Life!**

What do mercury, lead, chromium, copper, cadmium, and silver all have in common? That's right, they are all metals—heavy metals. They form colorful compounds, or *colorants*, that have been used since ancient times in paints, dyes, inks, and cosmetics. These metals have enriched and colored our lives. But if carelessly released into the environment, they are potentially toxic to living things.

### Challenge

Read about and discuss the historic use of toxic heavy metals and their risks and benefits to society.

Students testing the properties of different inks.

10,000 B.C. – rope is developed

## Source Reduction

### Copper: The Glory of Blue-Green!

Copper and its compounds have many beneficial uses. Copper plating helps protect metals from corrosion. Without wires made of copper metal, the long distance transmission of electricity would be too expensive. Copper compounds provide the rich blue-green color of turquoise jewelry. The ancient Egyptians discovered that when malachite, a copper ore, was powdered and mixed with natural oils, it made an excellent cosmetic. The greenish-blue color was the perfect complement to women's eyes and the rage of the times. Cleopatra probably used it for eye shadow!

Copper in very low concentrations is essential for human health. Yet in slightly higher concentrations, copper compounds are toxic to living things. Before the U.S. Pure Food and Drug Act of 1906 was enacted, bluestone, or copper sulfate, was added to pickles to make them look fresh and have a dark green color. It was once commonly poured into public swimming pools to control algae! Today, copper compounds are no longer used in food, cosmetics, or pools.

### Lead: From Pottery to the Color of the Queen's Face!

One of the densest of all metals is lead. For thousands of years potters used lead compounds for the colorful glazes on their pottery. These compounds are leached easily by mild acid solutions and can be transferred directly from dishes and containers to the food they hold. Today, in the United States pottery and ceramic items must pass rigorous tests to guarantee them to be virtually free of lead. But older ceramic items, and especially imported ones, should be tested for lead before using them for eating or drinking purposes.

## Source Reduction

Not only was lead used in glazes, it was also used in face powder, paints, inks, and even food coloring. In 16th century England, Queen Elizabeth I used a face powder made from white lead. In those days people did not bathe so the daily application of lead from the powder built up in her body, eventually leading to her death.

Until recently, lead made up nearly 50% of white house paint. Not only white, but bright red, orange, and yellow paints were also colored from a mixture of lead and chromium compounds. Inks also contained lead. Red lead was used in food, too. Less than 100 years ago this red colorant was added to candy and tea as a food coloring!

Lead has no known function or health benefit for humans. Its compounds are highly toxic. They are easily absorbed through the skin, the lungs, and the digestive system. Because the body cannot tell the difference between lead and calcium, lead accumulates in bones and can build up over time to potentially toxic levels. Young children are especially susceptible to the toxic effects of lead when they eat chips of paint from older buildings painted before the early 1980s.

Recently it was discovered that a new source of lead was entering the human diet. This time it came from the ink on plastic bread bags. It happened because people wanted to recycle the bags for storing other food. They turned the bags inside out and reused them. Of course, any slightly acidic material, such as fruit, could cause the lead to leach from the ink. Researchers estimated that the lead leached was nearly double the normal intake of lead for most people. The researchers concluded that lead ink on such bags should be prohibited as an "unnecessary risk to health."

## Source Reduction

### *Synthetic Dyes from Coal and Petroleum*

Both lead and copper compounds were regarded as natural colors. They were found in their natural state in the Earth as minerals, and they were used in much the same way as natural dyes extracted from plants. As we have seen, many of these natural mineral products were quite poisonous. However, there were no alternatives for these colorants until the discovery of petroleum and, with it, the discovery of synthetic colors and dyes.

In 1856, Sir William Henry Perkins of England synthesized the color mauve from coal tar oil. About three years later, with the discovery of oil in Pennsylvania, large amounts of petroleum became available to make other dyes and colorants. These synthetic dyes quickly proved to be superior to their plant and mineral counterparts that had served civilization for more than 20 centuries. By 1900, a total of about 80 dyes had been made. Over time, with testing and strict governmental controls, many of these coal- and petroleum-based dyes have become a part of the inks, dyes, and food colorings we use today.

The nature and the amounts of the colors that come into our lives have changed. Synthetic coal tar- and petroleum-based colorants have replaced many of the early mineral pigments in paints and inks that were toxic to humans. They were considered safe until further research found problems with some of them. However, some still remain. Today, all colorings must undergo rigorous tests before they are allowed to be used in the products and foods that we use daily. Some inks today are soy-based and are considered safer.

*Source Reduction*

# 45

## Introduction — Reducing Hazardous Waste

For many years people have been using dyes to color clothing, paint, cosmetics, and other products. Have you ever wondered how the beautiful colors of some dyes are produced? In this activity you will learn about the materials used to produce dyes and some of the problems in the disposal of these materials.

## Challenge

In this activity you will compare the properties of two inks and consider the tradeoffs involved in choosing one of them for printing a newspaper.

Inks are available in a wide variety of colors.

*Activity 45*

# 45

*Source Reduction*

## Materials

*For each group of four students:*

- One 30-mL dropping bottle of each of the following:
  - Water
  - 0.1 M potassium thiocyanate solution
  - 20,000 ppm iron (III) nitrate solution
  - 0.5 M hydrochloric acid (HCl) solution
  - 5% household ammonia solution
- One 15-mL bottle of 0.1 M potassium hexacyanoferrate (II) solution

*For each pair of students:*

- One SEPUP tray
- One piece of white paper
- One paper towel
- Two 1.3-cm x 15-cm pieces of newsprint
- One dropper
- One stirring stick

# Source Reduction

## Procedure

### Part One: Testing for Iron

**Safety**

Wear protective eyewear. Caution! the inks in this experiments may permanently stain your clothes.

1. Take out your SEPUP tray. Fill the large cups—A, B, and C—three-quarters full of tap water.

2. Now, add one drop of 20,000 ppm iron (III) nitrate solution to Cup A, Cup B, and Cup C. Stir.

3. Add 2 drops of household ammonia solution to Cup A. Stir. Clean the stirring stick on the paper towel.

4. Add 2 drops of 0.1 M potassium hexacyanoferrate (II) solution to Cup B. Stir. Clean the stirring stick.

5. Add 2 drops of potassium thiocyanate solution to Cup C. Stir.

6. In your Journal, prepare a data table to record the solutions and colors in each cup.

### Data Processing

1. What do all of the large cups have in common?

2. If the 20,000 ppm iron (III) nitrate solution represents a simulated toxic heavy metal, what concerns would you have about using these colored solutions for colorants in inks and paints?

3. What is the approximate iron concentration in Cup C? (Hint: See Activity 40.)

*Continued on next page→*

# 45

**Source Reduction**

**Procedure (continued)**

*Part Two: Making Your Own Inks*

Historically, the blue color you produced in Cup B has been used to make blue inks and dyes. It is called Prussian blue. In this part of the activity you will investigate two unknown blue ink samples. One ink represents Prussian blue and contains a simulated toxic heavy metal. The other ink represents modern replacements that have little or no heavy metal content. It will be your job to evaluate how useful each would be as a newspaper ink and recommend which ink to use.

1. Each pair of students in your group should test one of the ink samples—A or B. Decide which sample each pair will test.

2. Take your piece of newsprint to the stamp pad station you selected. Carefully apply the stamp to the ink and then press it to the paper. Repeat this three more times to produce a total of four equally spaced images on the piece of paper. See the example below.

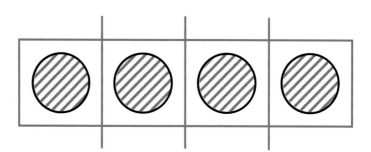

3. Cut the stamped pictures into four small, equal-sized pieces to fit the small cups of your SEPUP tray.

4. Each team will use one SEPUP tray to test their ink sample.

5. Put the pictures into small Cups 1-4. Wait 2 or 3 minutes for the ink to dry.

6. Add 6 drops of water to Cup 1.

7,000 B.C. – sun-dried bricks used for buildings

Activity 45

*Source Reduction*

7. Add 6 drops of bleach solution to Cup 2.

8. Add 6 drops of hydrochloric acid solution to Cup 3.

9. Use your dropper to add 6 drops of soap or detergent solution as assigned by your teacher to Cup 4. Wait 5 minutes.

10. Make a data table like the one below. Record your observations of the contents of each cup under the column for the ink sample that you are testing.

*Comparison of Two Unknown Inks*

| Cup | Substance | Effect on Ink A | Effect on Ink B |
|---|---|---|---|
| 1 | water | | |
| 2 | bleach | | |
| 3 | hydrochloric acid | | |
| 4 | soap/detergent | | |

11. Next, in your data table, record the test results obtained by the other team in your group.

12. Add a drop of hydrochloric acid solution to Cup 1 and Cup 4. Remember that Cup 3 already contains acid. Add a drop of potassium thiocyanate solution to Cup 1, Cup 3, and Cup 4 to test for the presence of our simulated heavy metal (iron).

13. Record the results of your tests in a data table like the one on the next page.

14. Clean up, as directed by your teacher.

*Activity 45*

## Source Reduction

*Results of Testing an Ink for Simulated Heavy Metals*

| Cup | Color of solution with hydrochloric acid and potassium thiocyanate | Heavy metal present? Yes or no? |
|---|---|---|
| 1 | | |
| 2 | | |
| 3 | | |
| 4 | | |

### Data Processing

1. Which ink contained the simulated heavy metal?

2. What is the purpose of the water in Cup 1?

3. What evidence do you have to prove that you tested two completely different inks?

4. To be useful for printing, inks must not be easily washed out by water. Which ink is the most resistant to water?

5. Printing inks must not lose their color quickly when exposed to light. The bleach solution was used to represent the effects of sunlight bleaching. Which ink is the most resistant to bleaching?

6. Some inks, over time, react with the acids that are found in many papers. Which ink was the most affected by acid?

6,000 B.C. – earliest known woven cloth

## Source Reduction

### Introduction

**The Green Dot**

Many local and national governmental agencies require their suppliers to use products that contain 25% recycled materials. Without these regulations, many recyclables would never be used because the raw materials that are used to make new ones would be cheaper. To make source reduction happen, governments in Europe require manufacturers to take back all their waste packaging. The following article will help you understand this program by showing you how a common product—an ink jet cartridge (used for some computer printers)—was redesigned to reduce the packaging and the added waste the manufacturer would be required to dispose of in a landfill or waste-to-energy incinerator.

### Challenge

After you read the article, think of ways to reduce the packing on some products you use.

#### Part One: Redesigning Packaging

European countries were the first to employ large scale incineration of their household waste. But they wondered why so many things they used couldn't be reused, reduced, or recycled. They asked, "Why do so many things need to be thrown 'away'?" One of the easiest areas to change (reduce) was packaging.

Laws were passed to reduce packaging waste by giving products the "green dot" environmental symbol of approval. Green dots told consumers they could return the used packaging to the place they bought it or to a collection site, and the producer would be responsible for its reuse or disposal.

# Source Reduction

Manufacturers soon got the message. They began to redesign their products in order to reduce the amount of packaging used. For example, liquid detergents that were packaged in large plastic containers were replaced with a highly concentrated form of detergent in smaller paper containers. These concentrates could now be diluted in the larger plastic container, just as orange juice is made from a concentrate. The large container could be reused over and over again.

What follows is a case study of how a producer of ink jet printers in the United States, Hewlett-Packard Company, completely redesigned its packaging for its ink cartridges in order to reduce the waste and obtain the green dot in Europe.

### Part Two: Packaging Ink Cartridges

Have you had a message from a computer lately? Computers can communicate with us through the monitor screen—and some can even talk! But to keep a written record of our work, we use a printer. An increasingly popular and inexpensive kind of printer, called an ink jet printer, sprays dots of ink to form images. These printers are used in homes across the United States.

Ink jet printers use a removable, single-use cartridge that fits inside the printer. It moves back and forth across the paper surface, spraying out a specially engineered ink. The ink creates a changing pattern of dots that can become any style of type.

The cartridge head contains a gold-plated computer chip that determines which small holes the ink will flow through. Tiny, heat-producing resistors are connected to each hole and to a supply of ink. The computer sends a message code to the printer head causing the resistors to instantly heat up and produce a small ink bubble. The bubble explodes through the hole and smashes into the piece of paper, forming a dot. The pattern of dots produced by this and other exploding holes firing ink missiles at the paper forms the letters we eventually see.

## Source Reduction

The ink cartridge is a marvel of modern engineering techniques. But like any high-technology device there are problems it may encounter during its shipping and storage before it ever reaches the consumer.

The inks in the cartridges are water based. They were designed to dry quickly, be nontoxic, and leave a lasting image on many different types of paper. Over time in storage, the water can evaporate, leaving a more concentrated ink behind—one that may clog the tiny holes in the cartridge's printer head. A second problem is transportation, causing the bumps, shakes, and drops the cartridge must endure on the way to the consumer. These may also damage the delicate printer head. The original package was designed to prevent evaporation and to protect the cartridges.

Originally the cartridge was packaged using seven separate components. (See the diagram below.)

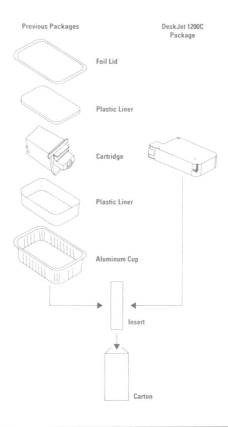

**Source Reduction**

Aluminum was chosen because it prevents water from evaporating from the cartridge. The plastic liners and carton helped cushion the shocks of transport. But when this product was introduced into Europe, it had to reduce its packaging to qualify for the green dot.

This became a formidable challenge to the designers. After considering the tradeoffs, both of cost and high quality for the consumer, the whole cartridge was redesigned. The aluminum barrier to prevent the ink from evaporating became an inside layer of the ink bag in the cartridge. Sturdy plastic parts were employed, and a carton with air cells (like the foam you studied) became a part of the packaging. What had been seven components, now became three. And Europe's gain became ours, too! The same cartridge is widely used here as well.

Manufacturers are now considering many ways to reduce the amount of packaging that eventually reaches landfills. You may remember the large packages compact discs (CDs) once came in or the large packages of laundry detergent. New designs reducing the amount of plastic in soft drink bottles are being used today. As new materials and products appear on the store shelves, they will always need some form of packaging. Companies will be continually challenged to provide minimal packaging that reduces waste in order to be environmentally responsible.

**Questions**

1. What are at least five purposes that packaging serves?

2. Think of a product that you use frequently and how it is packaged. Name the product, give three reasons you think it is packaged this way, and make two recommendations for reducing the amount of the packaging materials.

3. Is it possible to sell products with no packaging? Explain.

4,000 B.C. – kiln-dried bricks used — bronze and copper alloys

# Activity 46: Integrated Waste Management

## Introduction

### Comparing Waste Management Plans

Some communities do not rely on one specific method for all types of waste. They use an integrated approach to waste management. This method brings together several approaches from the waste hierarchy to create an overall plan to meet the needs of the community.

## Challenge

Your group will examine case studies of four counties across the United States that have developed waste management programs identified by outside groups as being outstanding.

Waste-to-energy facilities like this one convert chemical energy in waste to useful forms such as electricity.

# Integrated Waste Management

## Procedure

Each group member should choose a different case study to read. When you finish reading, answer the questions below in your Journal. Use the county name as the heading for your paper.

## Questions

1. What are the goals for managing waste in this county?
2. What methods are used to accomplish these goals?
3. What is unique about this plan? How does it take advantage of local conditions and contribute to protecting the quality of the environment?
4. What evidence is given about monitoring and safeguards to protect the environment from potentially toxic and hazardous substances?
5. What other information would you like to have about the county plan?

*Integrated Waste Management*

## Case Study 1
## Mecklenburg County, North Carolina, 1992

Mecklenburg County is a growing area, spurred by the economic development of its principal city, Charlotte. By the year 2006, the current population is expected to increase by 30%, and employment opportunities will increase by 50%.

In 1985, most of the county's waste was sent to landfills. County officials then began to develop a plan for integrated solid waste management. The plan calls for the county to recycle or reduce 40% of its solid waste, incinerate 30% in a waste-to-energy plant, and landfill the remaining 30% that could not be recycled, reduced, or incinerated by the year 2006. Since the plan was initiated, recycling activities have steadily increased, two bond issues have been passed to support waste-to-energy facilities and other solid waste management programs, and additional landfill areas are being sought.

Recycling and source reduction have reduced by 9% the quantity of waste disposed. The goal is to reach a 25% reduction and recycling rate by 1993. To accomplish this, landfill-user fees helped fund the county's $1.5 million recycling operations budget. Citizens were offered a reduction on their landfill fees if they brought in a set amount of recyclable material.

With the use of color-coded containers, curbside collection of aluminum, glass, newspaper, and plastic began in all towns and cities within the county. Now, over 74% of the homes participate in curbside recycling. Nearly one-fourth of all yard waste—leaves, grass, and other clippings—is recycled and composted. An extensive recycling program for commercial waste to recover office paper, corrugated cardboard, iron-containing metals, and scrap wood will begin soon.

In 1989, the county's first waste-to-energy incinerator began operation. Approximately 10% of the county's solid waste is treated in this incinerator. Steam from the facility is sold during the winter months to heat university campus buildings. Excess electricity is sold to utilities. Total annual energy earnings are estimated at $0.8 million, while the county must spend $2.4 million yearly to operate the plant.

*EPA Studies and Mecklenburg County Engineering*

2,000 B.C. – early glass

## Integrated Waste Management

### Case Study 2
### Pinellas County, Florida, 1994

Surrounded on three sides by the Gulf of Mexico and Tampa Bay, Pinellas County residents have a picturesque environment. But county solid waste authorities must also contend with nearly 25,000 new residents each year and a waste generation rate that is nearly 75% higher than the national average. To address their solid waste disposal needs, the county developed a four-part program that has helped reduce its reliance on landfills by 89% since 1983. The county's plan includes recycling, composting, a modern waste-to-energy plant, and an artificial reef program.

The recycling program that began in 1990 offers municipalities curbside pick-up of newspapers, glass, aluminum cans, and plastic beverage and milk containers (PETE and HDPE plastic). The program is subsidized by a county grant. Drop-off centers collect the same materials plus office paper, cardboard, aluminum foil, and used motor oil. The program has succeeded in recycling or composting 22% of the county's wastes.

The county's composting program, "Don't Bag It," encourages residents to leave grass clippings on their lawns. The clippings act as mulch and decompose, returning nutrients to the soil. Pamphlets relating tips on mowing, fertilizing, and home composting are also distributed. Finally, finished compost is available free at 17 locations throughout the county.

An important component of the county's solid waste management program is a modern waste-to-energy plant. The plant is run by 10% of the electricity generated; the other 90% is sold to Florida Power—enough to supply the needs of 45,000 homes. The plant is self-supporting. The sale of electricity and the fees charged to trash haulers who use the plant pay for its operation.

The county uses a unique disposal option as an alternative for some materials that would ordinarily be sent to landfills. A series of ten artificial reefs have been built off the coast using discarded construction materials, prefabricated structures, and old barges and ships. These reefs act as refuges for fish and provide protection against shore erosion caused by large storm waves.

*Adapted from* Case Studies in Integrated Waste Management.
*Used with permission by the Council on Plastics and Packaging and the Environment, 1994*

*Integrated Waste Management*

## Case Study 3
## Hennepin County, Minnesota, 1994

Hennepin County, which includes the city of Minneapolis and outlying communities, has developed an integrated approach to its waste management needs. Through county grants, widespread industry support, and hard work, the 46 cities that make up Hennepin County were able to reduce the amount of waste sent to landfills to 20%, down from 67% in 1989.

A permanent site for the collection of household hazardous waste was established. The county also began a program to collect batteries. There are more than 500 drop-off locations where batteries are collected and sent to a processor that reclaims the mercury and silver they contain.

In 1990, Hennepin County received the National Recycling Coalition's award for the best recycling program in the country. All cities within the county have curbside recycling programs. Major recyclables include glass containers, aluminum, tin (bi-metal) cans, newspaper, cardboard, and plastic bottles. There is one county drop-off center plus 30 local centers. The county provides financial assistance for programs that separate recyclable items into categories at their source of collection. The county spent approximately $6.5 million in 1990 to fund recycling programs. The composting of wastes is the responsibility of local cities. In the second half of 1990, 305,000 tons of municipal solid waste were recycled or composted, compared to 161,000 tons for the same period in 1989.

Hennepin is the sponsor of a waste-to-energy plant located near downtown Minneapolis. The plant began operation in late 1989 and burns an average of 1,000 tons of waste every day, generating enough electricity for 40,000 homes. Part of the fees charged to private and public waste haulers using the plant support other waste management programs. A second waste-to-energy plant is the Elk River refuse-derived fuel system in Anoka County. This operation began in 1989 and handles up to 1,500 tons of waste per day from five counties, including Hennepin. The waste is turned into fuel and then burned by a local power company to generate electricity. Hennepin County has a contract with the Elk River plant to burn 200–800 tons of waste per day, depending upon the seasonal flow of refuse.

Presently, 53% of Hennepin County's waste is processed by the waste-to-energy plants. These plants use dry scrubbers and baghouses as pollution control devices. They operate under the limits set by the State Pollution Control Agency.

*Adapted from* Case Studies in Integrated Waste Management.
*Used with permission by the Council on Plastics and Packaging and the Environment, 1994*

1,000 B.C. – fabric dyes used

# Integrated Waste Management

## Case Study 4
## Marion County, Oregon, 1994

Like many communities, Marion County had, for years, been using landfills for all of its waste disposal. Faced with the closure of the Brown's Island landfill in 1974, Marion County's 210,000 residents were left without an immediate disposal method. Through a comprehensive waste management program, the county was able to reduce the amount of waste sent to landfills from 100% in 1974 to just 7% today.

Oregon was the first state to pass bottle deposit legislation in 1971. The law places a deposit on all soft drink and beer containers and requires customers to return the containers to receive a refund. Oregon also has mandatory recycling for all cities with a population of more than 4,000 residents. The county went a step further in 1986, when it began a curbside recycling program. It requires trash haulers to pick up recyclables (glass, newspaper, aluminum, tin cans, corrugated cardboard, and used motor oil) on a weekly basis in urban areas and by an on-call basis in rural parts. There are also 20 drop-off locations in the county that accept these same materials along with grayboard, magazines, and PETE and HDPE plastic (recycling code 1 or 2). The program is paid for by the disposal and collection fees paid by the haulers. Haulers also pay for the drop-off program. A decline in the scrap value of many recyclables has led to a $90,000 deficit for the recycling program. Currently 25% of Marion County's municipal solid waste is recycled. Yard waste accounts for more than 20% of the county's municipal solid waste. There are four yard-waste composting facilities operating in the county. The county charges $55 per ton, and the two private facilities charge $40 per ton. This pays for the operation of the program. Approximately 10% of Marion County's yard waste is composted.

Marion County is the sponsor of a mass burn (all municipal wastes collected are burned together) waste-to-energy incinerator. The plant was completed in 1986 at a cost of $47.5 million. The plant can process 170,000 tons per year. The plant's 13.1-megawatt generator produces 11 megawatts of electricity that are sold to a local power company for approximately $4 million a year. The pollution control devices for the plant are dry gas scrubbers and fabric filter baghouses. Ash from the plant is sent to a special landfill. Today, 66% of Marion County's municipal solid waste is burned to recover energy.

*Adapted from* Case Studies in Integrated Waste Management.
*Used with permission by the Council on Plastics and Packaging and Environment, 1994*

# Integrated Waste Management

## Introduction

### Preparing an Integrated Waste Management Plan

Imagine that the solid waste landfill site owned and operated by your community will be full in four to five years at the current rate of waste disposal. This is not enough time to locate a new site and build a landfill that meets state and federal environmental regulations. In order to extend the life of the landfill and allow time for a new landfill to be built, local government has mandated a 25% reduction in the amount of waste sent to the landfill.

## Challenge

> Prepare an integrated waste management plan that will reduce the amount of waste sent to the landfill and extend the life of the landfill.

## Additional Information

To accomplish the challenge you should consider using a combination of waste management methods. You should also have the following goal for your solution: All wastes should be reduced in volume and toxic content, and useful energy and materials should be extracted prior to their final disposal. Remember, even when the new landfill is built, it won't last forever, and you want your community's long-term waste management plan to be safe and effective.

Past recycling efforts in the city have been sporadic and small in scale. There is no curbside pickup and there are no neighborhood collection sites. The city does not operate a waste incinerator. The percent by volume and weight of the city's solid waste is provided in the pie charts on the next page. Keep in mind that almost one-third of the paper in the landfill is newsprint.

200 B.C. – concrete    150 B.C. – paper

Activity 46

## Integrated Waste Management

### Composition of Municipal Solid Waste

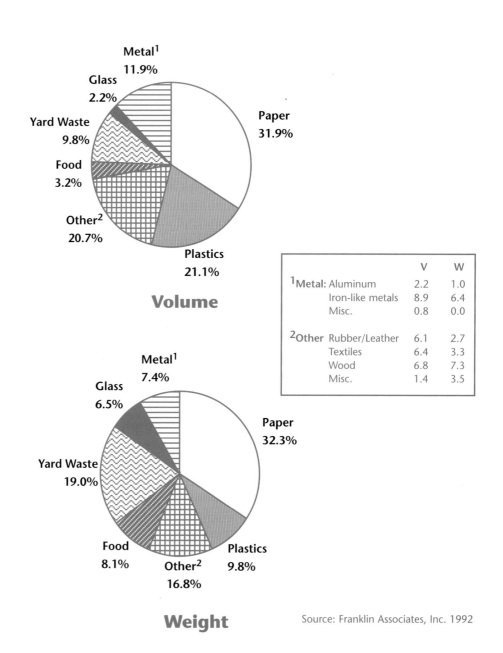

Source: Franklin Associates, Inc. 1992

A.D. 250 – use of directional compass

A.D. 200 – porcelain

Activity 46

*Integrated Waste Management*

## Procedure

1. Working in groups of four, develop an integrated waste management plan for the community. You may find it helpful to consider each category of waste described on the pie graphs on the previous page. Work with your group to identify at least five things that you think should be included in the waste management plan.

*Managing Household Wastes*

| Waste Category | Waste Management Methods | Reason for Choice |
|---|---|---|
| Paper | | |
| Plastics | | |
| Glass | | |
| Yard Wastes | | |
| Metals | | |

2. Decide how your group will present its solution to the class. In your presentation, explain why you are convinced that your plan will accomplish the 25% reduction in the amount of waste going to the landfill for disposal.

3. In your Journal explain the highlights of your group's plan and how the plan will reduce the waste going to the landfill by 25%.

A.D. 600 – books printed
A.D. 650 – cotton introduced
A.D. 700 – sulfuric and nitric acid invented
A.D. 850 – primitive use of gunpowder

Activity 46

# Materials and Applications Timeline

The bottom line on each page of this part of your Student Book is a timeline for materials and their selected applications.

Each page represents a period of 500 years, so the timeline begins 73,500 years B.C. These two pages represent the last millenium, from A.D. 1000 to the present.

Important dates in materials and their applications are listed. Many of the dates are approximate and depend on written history. It may be that in some cases the materials were used long before someone decided to write about them.

The thick line above the timeline represents the number of people living on Earth. Of course, that number can only be estimated. The population at the beginning of this timeline was probably about 2 million people.

- 6 billion people
- 5.5 billion people
- 5 billion people
- 4.5 billion people
- 4 billion people
- 3.5 billion people
- 3 billion people
- 2.5 billion people
- 2 billion people
- 1.5 billion people
- 1 billion people
- 500 million people

1300 — quill and ink pens
1450 — first printed Bible
1500 — black lead pencils used
1300 — eyeglasses invented

human population

A.D. 1000  1100  1200  1300  1400  1500

146

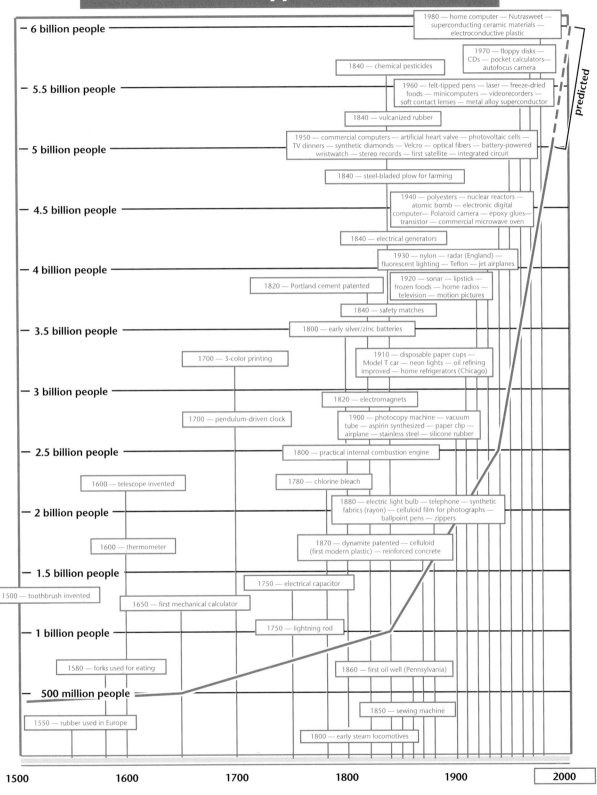

# Part Three

# Energy

# Part 3
# Table of Contents

*Introduction* .................................................................................................. 1

47. Batteries: Energy and Disposal ...................................................... 2

48. Electroplating .................................................................................. 15

49. Electrical Appliance Survey ........................................................... 19

50. Investigating Energy Transfer ....................................................... 27

51. Developing an Energy Savings Plan ............................................. 32

52. Ice and Energy Transfer ................................................................. 37

53. Quantifying Energy: Calorimetry ................................................. 45

54. Efficiency: Energy Changes and Waste ....................................... 56

55. Electrical Energy: Sources and Transmission ............................. 65

56. Energy from the Sun ...................................................................... 69

57. Controlling Radiant Energy Transfer .......................................... 78

58. Designing an Energy-Efficient Car .............................................. 80

# Part 3 Energy

## Introduction

What is energy? How do we use it? Where does it come from? Are there enough energy resources for us to continue using energy at the same rate into the 21st century? How can we decrease the amount of pollution that results from our use of energy? These are some of the questions you will investigate in the next part of *Issues, Evidence and You*.

In Part Two of the course, you discussed the use of paper or plastic bags to carry groceries. Some of you may have been concerned about the fact that plastic bags are made out of petroleum, a nonrenewable resource. Did you know that it can take more oil to make a paper bag than a plastic bag? How can that be? Although paper bags are not made directly from oil, oil may have been used to produce the energy used by the paper mill. Since paper bags take more energy to make than plastic bags, the amount of oil used to make a paper bag may actually be more than was used to provide the energy and raw material for a plastic bag! But you would never think of that unless you were paying attention to the energy costs of making the bags. Any attempt to make decisions about the best materials or processes for reducing environmental impact must consider energy as one factor in the decision.

In Part Three of the course, you will learn more about energy sources and how we use these sources to produce energy in the forms we need for the many things we do. You will investigate energy chains and energy efficiency. Most importantly, you will relate what you learn about energy to your own life and think about yourself as an energy consumer.

When you complete this part of the course, you will be able to use what you learn about energy and energy sources and what you have learned about materials to come up with your own plans for a more energy-efficient future.

# Activity 47
# Batteries: Energy and Disposal

## Introduction

### Batteries: A Closer Look

Think of a battery as an energy converter. It is able to convert chemical energy into electrical energy. Look at the accompanying diagram of a battery and read about the various parts and how they work.

## Challenge

Your challenge is to gain a better understanding of how a battery works.

**D-Cell Carbon-Zinc Battery**

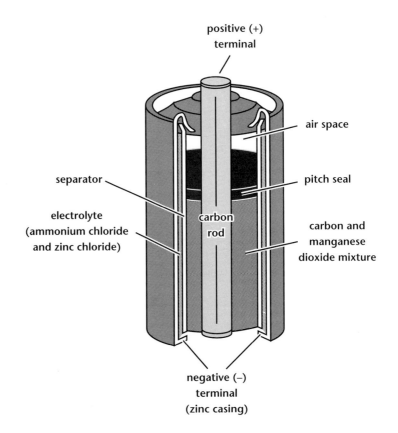

# Batteries: Energy and Disposal

A common flashlight battery, also called an electric cell, is pictured on the opposite page. All batteries contain a liquid, paste, or solid material that electricity can flow through. This material is called an *electrolyte*. The electrolyte in the D-cell battery is a moist paste of ammonium chloride and zinc chloride. The cell is sealed at the top (pitch seal) to prevent moisture from evaporating.

Batteries also have two terminals, a positive terminal and a negative terminal. The D-cell's positive terminal is the carbon rod in the center and the carbon and manganese dioxide mixture located around it. The negative terminal is the zinc outside shell of the battery.

When you put a battery in a flashlight and turn the light on, you are producing an external circuit or path for electron flow. A chemical change takes place inside the battery. The zinc (negative terminal) starts to react with the moist paste, leaving an excess of electrons on this terminal. The electrons flow through the external circuit back to the carbon rod (positive terminal) which accepts them. This flow of electrons is the electrical current which passes through the flashlight bulb and produces the light.

In the carbon-zinc battery, this chemical change cannot be reversed. The battery dies when the reactants are used up. This type of battery is called a primary cell. Other types of batteries called secondary cells are rechargeable. The lead-acid batteries in motorcycles and automobiles are examples of secondary cells. These two types of batteries are pictured on the next page.

### Question

Explain in your own words why a battery "dies."

# 47 Batteries: Energy and Disposal

## Batteries and the Environment

### Single Use—Not Rechargeable (Primary Cells)

**Zinc-carbon cell batteries**
Classic, Heavy Duty, Super Heavy Duty Plus
- zinc (a toxic heavy metal)
- zinc chloride (toxic by ingestion)

**Alkaline batteries**
Energizer, Duracell, Supralife
- potassium hydroxide solution (a very strong base)
- zinc metal

**Lithium batteries**
Ultralife, Lithion, Procell
- lithium metal (causes fires and explodes with contact with water)
- iron dissulfide (corrodes metals, produces hydrogen sulfide gas—more poisonous than gas at the same concentration)

**Mercury batteries**
hearing aids
- zinc metal
- mercury oxide (heavy metal oxide—accumulative poison)
- potassium hydroxide

**Silver oxide batteries**
watch, calculator
- silver oxide (toxic heavy metal)

### Rechargeable Batteries (Secondary Cells)

**Lead-acid batteries**
car, boat, motorcycle
- lead plates (toxic heavy metal)
- lead dioxide plates (toxic heavy metal compound)
- concentrated sulfuric acid

**Nickel-cadmium batteries**
rechargeable, Enercell
- nickel oxide (toxic heavy metal oxide)
- cadmium metal (toxic heavy metal)
- potassium hydroxide solution

### Hazard Categories
Toxic: poisonous
Corrosive: can burn through metals and flesh, acids and bases
Flammable: catches fire, explodes
Reactive: combines with other materials to form poisonous gases, fires

Activity 47

# Batteries: Energy and Disposal

## Introduction

### Chemical Batteries

Batteries are objects that contain materials in which chemical changes take place. You will construct some simple chemical batteries and explore how different metals that take part in a chemical reaction produce electrical energy. You will use a small motor to detect how much energy is produced.

## Challenge

Test different combinations of metals to see which ones produce the most energy. You will also investigate what happens to the energy produced as the two metals are brought closer together. And, if time permits, explore how changing the concentration of the reactants affects the energy produced.

This is a lead-acid battery that is used in automobiles.

# 47 Batteries: Energy and Disposal

## Materials

*For each group of four students:*

- Two strips of each of the following metals:
  - Copper
  - Zinc
  - Iron
- Two 3-inch strips of magnesium
- One 30-mL dropping bottle of 3% hydrogen peroxide solution
- One small piece of steel wool

*For each group of two students:*

- One SEPUP wet cell chamber
- Four packages of table salt
- One stirring stick
- One 30-mL graduated cup
- Two jumper leads—one red and one black
- One paper towel
- One 9-ounce plastic cup
- One electric motor
- Masking tape

Activity 47

# Batteries: Energy and Disposal

## Procedure

### Part One: Effects of Different Metals

The first goal of this investigation is to discover which two metals produce the most electricity when set up according to the directions below. Be sure to try all possible combinations. Use the list of combinations you copied from the class discussion to guide your testing. As you test each combination, share your observations with the other members of your group and record the results in your Journal.

**Safety**

Be sure to use safety eyewear during this investigation. Hydrogen peroxide can discolor clothing.

Two members of your group should test the three combinations of metals with copper. The other two group members should test the other three combinations.

**Note:** It is extremely important to dry the metals on a paper towel and clean both sides before using them for the next test!

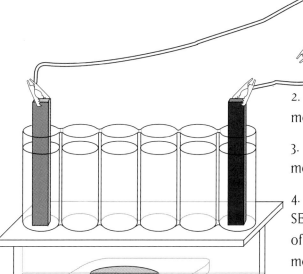

1. Add water to near the top of the SEPUP wet cell and stir in 4 packages of salt. Add 25 drops of hydrogen peroxide.

2. Put a small piece of masking tape on the motor shaft.

3. Pick two of the metals. Clip a wire onto each metal and then clip the other ends to the motor.

4. Lower the pieces of metal into the slots of the SEPUP wet cell as shown. Leave a little sticking out of the cells. Observe how fast and how long the motor spins. If the motor still spins after 2 minutes, stop timing. Remove the two metal pieces. Dry them, and then shine them with a piece of steel wool.

*Continued on next page→*

**Procedure (continued)**

### Part Two: Effect of Distance Between Metals

As a group, determine what happens to the energy produced by the battery when two metals are placed closer to each other. For this test use the zinc and iron strips as your two metals. Place them at opposite ends. Gradually move the zinc strip closer to the iron strip. Make a table in your Journal to record the results.

### Part Three: Effect of Reversing Connections

Explore what happens to the direction the motor turns when you reverse the connections on a pair of metals. Use the zinc-copper combination of metals to explore this. What does this tell you about the direction (path) that electrons are flowing in the circuit?

### Part Four: Effect of Metal Surface Area

Use the zinc and copper combination to explore what happens to the energy produced by the battery as you gradually remove one of the metal strips from the battery solution.

### Data Processing

1. Construct a data table in your Journal to record your observations.

2. From your observations, which combination of metals produced the most electrical energy? The least energy? What is your evidence?

3. How does placing the metal strips closer to each other affect the speed (the amount of energy produced) of the motor? Propose an explanation for this.

4. How does the amount of the metal exposed to the solution change the energy produced?

5. If you were given copper, silver, and nickel to make a battery, what kind of experiment could you do to find out which combination of two of these metals would produce the most energy? Explain how you came up with your design.

# Batteries: Energy and Disposal

## Going Further

As time permits, investigate one of the following questions. First, design an investigation and write up the procedure in your Journal. After your teacher approves your design, do the investigation and record your results. Write a one- or two-paragraph summary of what you did and what you can conclude from your investigation.

1. How does changing the concentration of the salt or hydrogen peroxide affect the amount of energy produced?

2. What other combinations of metals can be used to power the motor?

# 47
## Batteries: Energy and Disposal

### Introduction
**You'll Get a Charge Out of These!**

Batteries and electricity are related to each other. You will read about how batteries were discovered and learn about how many batteries you would need to provide electricity for different events.

### Challenge

What are batteries? Who first discovered them? How many batteries would it take to keep your heart beating? Read on to find answers to these questions.

Life without a Walkman, Gameboy, or a flashlight may be hard to imagine. Yet, a little less than 200 years ago, the device that powers each of these was unknown.

Eighteenth-century scientists like Benjamin Franklin and Luigi Galvani provided early clues to electricity's potential to help people. For example, Franklin tied a key to a kite string to prove that lightning was actually electricity. The experiments of Italian scientist Luigi Galvani led to ideas that helped develop modern batteries. Galvani was a professor of anatomy at the University of Bologna in 1786. While dissecting a frog, he noticed that when the frog's muscles were touched with two strips of different metals, zinc and copper, the legs twitched—they actually moved! He also noticed that the sparks from a nearby electric machine made the frog's legs twitch. Galvani explored this further and called the cause of this movement "animal electricity." Although this idea of animal electricity was discredited, Galvani's work lead to the discovery of electrical currents, whose action is described by the verb, *galvanize*.

# Batteries: Energy and Disposal

Later, in 1800, Count Alessandro Volta of Italy invented the battery—the first source of electricity that could produce a current. This invention made the development of the telegraph and telephone possible.

The chart below will give you an idea of how electrical power in kilowatt-hours is related to some common events. It will also tell you how many D-cell batteries would be needed to produce the same amount of energy.

*Battery Power*

| Kilowatt-hours | What Produces the Power | Equivalent No. of D-cell Batteries |
|---|---|---|
| 1,000,000 | Energy produced by a large electrical generating plant per hour | 50,000,000 |
| 100,000 | Energy consumed per person in the U.S. each year | 5,000,000 |
| 10,000 | Energy needed to refine one ton of steel | 500,000 |
| 1,000 | | |
| 100 | | |
| 10 | Car trip of 40 kilometers (24 miles) | 500 |
| 1.0 | Average person's daily intake of food energy | 50 |
| 0.1 | One average person's physical work in one day | 5 |
| 0.01 | Energy produced per hour by a flashlight | 0.5 |
| 0.001 | | |
| 0.0001 | | |
| 0.00001 | Energy produced per hour by the human heart | 0.0005 |

**A battery is actually an energy converter.** It is able to convert chemical energy into electrical energy. In the following investigation, you will explore how different combinations of metals affect the energy a battery can produce.

# 47 Batteries: Energy and Disposal

## Introduction

### Which Battery to Choose?

Each year in the United States, close to 2 billion batteries are thrown away. Think about that number—2 billion. To reach that number, every one of the 250 million people in the U.S. would have to throw away just eight batteries a year.

## Challenge

How would you select a battery to use? Compare the costs, both economic and environmental, to help answer this question.

To give you an idea of why so many batteries are discarded, let's consider a simple example. Imagine one D-cell battery can be used to run your portable radio (a large radio requires more like 4 to 8 batteries!). Look at the chart below to see how many changes of D-cell batteries would be needed to run your radio for 500 hours. Consider the costs. See Question 1 on page 14.

*Comparing Different Types of D-Cell Batteries*

| Battery Type | Cost per Battery | Hours Each Battery Lasts | No. Needed for 500 Hours Usage | Total Cost to You |
|---|---|---|---|---|
| Alkaline | $1.30 | 20.0 | 25 | $26.00 |
| Heavy Duty | $0.60 | 6.5 | 77 | $46.20 |
| General Purpose | $0.37 | 2.2 | 227 | $83.99 |
| Rechargeable | $3.40 | 2.5 | 1 | $20.62* |

*includes the cost of the battery charger and electricity for 200 recharges

*Activity 47*

# Batteries: Energy and Disposal

Was your decision based on cost? Was it based on convenience—how many times would you need to purchase batteries or replace them; or how long would your radio run before you would have to replace the battery? What about batteries as wastes? How might they affect the environment?

Many batteries end up in municipal landfills and incinerators and pose a potential hazard to the environment. Until recently, most non-rechargeable batteries contained mercury to help the cell perform well over long periods of time. Rechargeable batteries, while they do not contain mercury, do contain nickel and cadmium. As you recall, all of these metals are classified as toxic heavy metals.

For example, in 1985, based on the total number of batteries sold and probably discarded, 46,553 pounds of mercury, 1,170 pounds of nickel, and 796 pounds of cadmium were disposed as battery waste. Due to increased regulation of the disposal of heavy metals and the realization of the potential hazards of these metals (especially mercury) to the environment, battery manufacturers adopted programs to reduce the use of mercury in the manufacture of batteries. By the early 1990s, the mercury content of batteries was greatly reduced—from 21,410 pounds in 1990 to 1,500 pounds today. However, due to the popularity of rechargeable nickel-cadmium batteries, the number of these batteries sold today has dramatically increased. As a result, the amount of cadmium used in batteries has increased to 13,800 pounds, up from 796 pounds in 1985. This means that now there is a potential for more cadmium to enter the environment. This is one of the tradeoffs involved in the increased use of rechargeable batteries. Thus, in response to laws, the urging of environmental groups, and their own concern about the environment, battery manufacturers have begun an aggressive effort to collect and recycle nickel-cadmium batteries, similar to programs begun earlier to collect hearing aid mercury batteries. Some states, like New Jersey, have passed tough battery recycling laws and have mandated collection centers for all batteries.

# Batteries: Energy and Disposal

**Questions**

1. Think carefully about which battery you would use in the radio described in the first paragraph. What are your reasons for selecting it? Write these in your Journal.

2. What kind of plan does your community have for recycling used batteries?

3. Earlier you read that about 2 billion batteries are discarded each year. Because all of these batteries contain small amounts of heavy (toxic) metal, they increase the risks of contaminating the environment and exposing people to a greater chance of getting sick from the metals.

    a) What kinds of information would you need to assess the risk of disease from heavy metals in batteries?

    b) Assuming that the risk is too high, what are some of the ways you could reduce the risk? List as many ways of reducing the risk as you can.

    c) Now that you have a list of possible methods of reducing the risk, you need to make a decision about what action(s) you will take. What are some of the issues that you need to consider in order to decide what to do to reduce the risk?

# Activity 48

## Electroplating

### Introduction

#### Copper Plating

There are many ways to help prevent the corrosion of iron. It can be painted, covered with grease or wax, or even coated with a less corrosive metal. In electroplating, a base metal (in this case, iron) is covered with a thin layer of another metal (copper) to prevent corrosion and/or provide a more attractive finish.

### Challenge

Investigate the use of energy in electroplating.

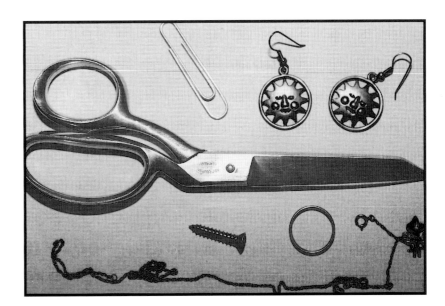

Metal plating can prevent corrosion and/or provide a more attractive finish.

# 48

## Electroplating

### Materials

*For each group of four students:*

- One paper clip
- One 180-mL bottle of acidified 0.25 M copper sulfate plating solution
- One 9-volt battery
- One battery harness with light bulb
- One graduated container
- One small copper strip

### Procedure

**Safety**

Wear protective eyewear. The copper sulfate plating solution is toxic and corrosive. Avoid contact with skin and eyes. Wash any exposed area with water for 2–3 minutes. Some people may have an allergic reaction to the copper sulfate solution, causing itching and redness to the affected area for a short time.

1. Attach your bulb with the battery clips to your 9-volt battery. Touch the two clips at the end together and notice how brightly the bulb shines. *To avoid wearing out the battery, do not allow the clips to continue to touch each other.*

2. Make a data table in your Journal. You will need to record the appearance of the copper strip, paper clip, and copper sulfate solution both now and after plating.

3. Bend a strip of copper metal in the shape of a "J." Hook the end of the strip over the side of the container, as shown in the diagram. Clip the red wire to the strip of copper, as shown.

4. Place a paper clip in the container, so that it is clipped to the sides of the cup and one part of it is in the solution. Clip the black wire to the top of the paper clip.

## Electroplating

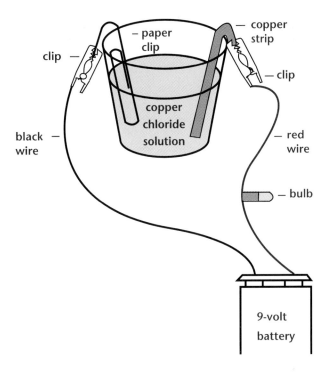

5. Add 25 mL of copper sulfate plating solution to the graduated container. Do not get plating solution on the clips of the battery harness. Avoid any skin contact!

6. Wait for a few minutes. Disconnect the battery clips. Remove the paper clip and copper strip and allow them to dry on the paper towel.

7. Record your observations of the paper clip, copper strip, and plating solution (copper sulfate) in the data table in your Journal.

8. Carefully pour the plating solution from your graduated container back into the container marked "Used Copper Sulfate Solution."

9. Rinse and dry your container, the paper clip, and the strip of copper.

# Electroplating

 **Data Processing**

1. What happened to the paper clip? What seems to be coating it?

2. Which side of the battery are the copper ions attracted to (positive or negative)? What evidence from your observations supports your answer?

3. As a group, discuss possible methods for disposal of the used copper sulfate solution. Write a paragraph in your Journal summarizing the ideas you have come up with and the advantages and disadvantages of each idea.

4. Describe how metal plating prevents corrosion.

# Activity 49
# Electrical Appliance Survey

## Introduction

### Appliances in 1902

You will conduct a survey of the appliances that you, your parents, and your grandparents have used over the generations. Before beginning the survey, here are some illustrations of appliances from the Sears, Roebuck and Co. catalogue of 1902.

## Challenge

Try to identify all of the objects shown on the next two pages. What kinds of energy were required in the home in 1902?

### Question

Judging from these appliances, what do you think was the main source of home energy in 1902?

# 49

## Electrical Appliance Survey

## Electrical Appliance Survey

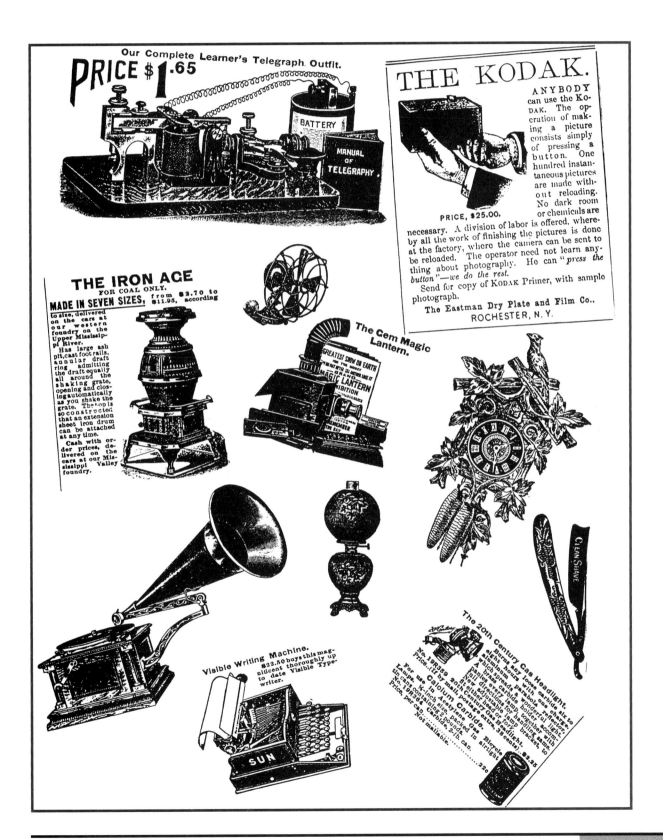

# 49

*Electrical Appliance Survey*

**Introduction** — **Utility Ads: Then and Now**

You will read three ads about energy: two of them appeared over 30 years ago and the other ad appeared in 1993. Study them to learn how people's attitudes about energy have changed.

**Challenge**

Your challenge is to find out how people felt about energy at the time each of the ads was written.

**Questions**

1. What is the common message in the two old ads? Give at least two specific examples of this message.
2. What is the message in the 1993 ad? Give one example.

*Electrical Appliance Survey*

## How many hands should a homemaker have?

ONLY TWO — if she makes full use of the many helping hands electricity has to offer!

Electricity will wash and dry her clothes... power the vacuum cleaner ... tell time... heat the water... mix a cake... dry her hair... brush her teeth... entertain the family... cook meals quickly and cleanly... even wash the dishes. And all for just pennies a day!

Small wonder we say electricity is by far the biggest bargain in your family budget. The more you use... the more time you have for your family!

LIVE BETTER ELECTRICALLY

### "Daddy says that's our handy man"

Daddy's right, of course. Home just wouldn't be as pleasant and comfortable without handy Reddy Kilowatt and the many conveniences made possible by electricity.

Today low-cost electricity lets us live better than the kings of yesterday. Most of the conveniences we now take for granted would have been luxuries a generation ago.

The American family is now using about twice as much electricity than 10 years ago. Yet in this same period the average residential cost per unit (a kilowatt-hour) has dropped 10%.

Other costs of living go up, but you're *living better electrically*, for less!

*Your Electrical Servant*
*Reddy Kilowatt*

*Live Better...Electrically*

®Reddy Kilowatt used with permission of The Reddy Corporation International.

# Electrical Appliance Survey

## ENLIGHTENING IDEAS FOR SAVING ENERGY

**You May Qualify To Have A Variety Of Energy-Saving Devices Installed In Your Home Free.**

NYSEG has some brilliant ideas to make your home more energy-efficient. That means a lower energy bill for you and a stronger, more beautiful New York for all of us. These are the free energy-saving measures NYSEG will help you take:

Replacing incandescent bulbs with high-efficiency lighting.

Installing automatic set-back thermostats in homes with central air conditioning and central electric heating.

Installing low-flow shower heads and faucet aerators to reduce hot water consumption.

Insulating hot water pipes to save heat and reduce your water heater's workload.

Wrapping the water heater and/or reducing the water heater thermostat setting to a comfortable but more energy-efficient level.

To find out if you qualify for these free home improvements, call your local NYSEG office. We'll shed some light on how you can make your home more energy efficient.

New York State Electric & Gas Corporation
**NYSEG**
*Good people. Good service.*®

# Electrical Appliance Survey

## Introduction

### Electrical Appliances: Then and Now Survey

By conducting this survey, you will collect data about the number of appliances you have in your home today, compared with the number of appliances your parents or guardians had at home when they were your age. The data will show any changes in the use of appliances and energy.

## Challenge

Take a look at the appliances listed on the survey distributed by your teacher. Guess how many you have in your home right now. Now conduct the survey and find out!

# Electrical Appliance Survey

## Procedure

1. In Column A, write the number (how many) of each kind of appliance you have in your home today. If you have an appliance that is not on the list, write its name on one of the blank lines.

2. Next, have a parent, or an adult you live with, fill in Column B with the *number* of each kind of appliance in the home *when he or she was your age*. If you can, ask a grandparent, an older neighbor, or other senior citizen to fill in Column C for the *number* of each kind of appliance in the home *when he or she was your age*.

3. When you have numbers filled in for Columns A, B, and C, add the numbers in each column to get the *total* number of appliances in the column for each category. Write the *totals* in each of the boxes.

4. Get a grand total by adding together the totals in each category. Fill in the grand totals.

## Questions

1. What is the total number of appliances that you have in your home today? How many appliances did your parent or guardian have when he or she was your age?

2. If your grandparent or an older neighbor was available to survey, how does the number of appliances in your house today compare with the number reported from this third generation?

3. What might be some reasons for the changes in the number of appliances over two or three generations? Which category of your survey had the largest changes?

Activity 49

# Activity 50: Investigating Energy Transfer

## Introduction

### House of Cards

It takes energy to construct a building. Where does the energy go? You will investigate that question in this activity. Working as a group, you have 10 minutes to build a house, up to five stories high, out of cards. The house must be able to support a foam drinking cup on top. The house that has the greatest number of stories and uses the fewest number of cards (therefore stores the least amount of energy) is considered the most successful.

## Challenge

Use the least amount of stored energy to build a five-story "House of Cards" that can support a foam cup.

## Materials

*For each group of four students:*

- 80 3x5 index cards:
    - 25 labeled #1 (6 folded)
    - 20 labeled #2 (3 folded)
    - 15 labeled #3 (1 folded)
    - 10 labeled #4
    - 10 labeled #5
- One foam cup

# Investigating Energy Transfer

## Procedure

### The Rules

1. Each group may build only *one* card house.
2. The structure must be freestanding and support the cup on top.
3. The cards may not be folded, except for the 10 prefolded ones.
4. The cards can't be ripped or cut.
5. Each "story" is made up of its walls and the ceiling above it.
6. Numbered cards correspond to each story and may be used only for that story.
7. You are allowed only 10 minutes for official construction. When time is up, the structure must stand, with the cup on top, without collapsing for at least 1 minute.
8. When directed by the teacher, your group will release the energy now stored in the structure and calculate the number of "energy units" the house contained. Use the directions on the next page for calculating energy units.

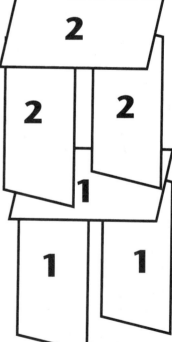

Activity 50

# Investigating Energy Transfer

 **Data Processing**

### Calculating Energy Costs

Adding up the energy that was stored in your finished structure is easy. Because each card is the same size and weight (whether used as a wall or a ceiling panel), the energy it takes to lift one is the same for each. It's also easy to understand that lifting one twice as high takes twice the energy, and three times as high takes three times the energy, etc.

Energy cost per card:

| | | |
|---|---|---|
| first story | = | 1 energy unit |
| second story | = | 2 energy units |
| third story | = | 3 energy units |
| fourth story | = | 4 energy units |
| fifth story | = | 5 energy units |

Use a chart like the following one, or one your group designs, to help you calculate the stored energy:

*Calculating Stored Energy Units*

| Story | Number of Cards | Energy Unit per Card | Total Energy Units per Floor |
|---|---|---|---|
| First story | | times 1 | = |
| Second story | | times 2 | = |
| Third story | | times 3 | = |
| Fourth story | | times 4 | = |
| Fifth story | | times 5 | = |
| **Total energy units in building** | | | |

Activity 50

# 50 Investigating Energy Transfer

### Introduction

**Drive a Nail**

In this activity, you will design experiments to help you understand the differences between potential and kinetic energy. You will investigate the variables that determine how far a nail is driven into a foam block by a falling weight.

### Challenge

Design and carry out investigations to determine the effect of height and mass (weight) on the energy transferred to a falling object.

### Materials

*For each group of four students:*

- Two plastic tubes (one tall and one short)
- Two steel weights (one large and one small)
- Two aluminum weights (one large and one small)
- Two steel nails
- Two aluminum nails
- Two blocks of foam

# Investigating Energy Transfer

## Procedure

**Safety**

Use care when dropping the cylinders; do not throw them.

To do this experiment, use the materials your teacher provides and follow the safety precautions demonstrated in the basic test procedure that your teacher modeled for you. Keep in mind that you have two plastic tubes—one twice as tall as the other—that will help you control the drop height. You also have both steel and aluminum weights. You are provided with two of each kind—one twice as heavy as the other—to make it easy to control the weight of the object dropped and to investigate the effect of weight on the amount of energy transferred to the nail.

The actual testing will be done in the next science class. You should take the remainder of this period to develop a group plan for that test. Keep in mind: You want to carry out a controlled experiment that will determine the effect of drop height and weight on driving a nail completely into the foam. (Two of you might test the aluminum weights and two the steel weights.) Be sure you are all clear on exactly how to drop the weights from the top of the tubes; any variation will likely cause mixed results. Making a group data table now will help your group think about what needs to be tested and who's going to do what.

Before beginning the experimental work, predict your answers to the following questions in your Journal and explain your predictions:

1. Which combination of tube height (tall or short) and metal weight (steel or aluminum/large or small) will drive the nail deepest into the foam block? Explain how you arrived at your answer.

2. Which combination of tube height and metal weight will require the greatest number of drops to pound the nail completely into the foam block? Explain how you arrived at your answer.

### Extension

Ask your teacher for a different type of nail. Determine if there are any differences in the energy it takes to drive it into the foam block.

# Activity 51: Developing an Energy Savings Plan

## Introduction

In this activity you will take another look at your top ten appliances lists. You will determine how much electricity each appliance uses in a year and calculate the total cost of the electricity.

## Challenge

Come up with a plan to save 20% of the cost of electricity on one of your top ten lists.

## Procedure

1. Locate the two top ten lists that you recorded in your Journal for Activity 49, "Electric Appliances: Then and Now Survey."

2. Add four columns with these titles to each list:

| Power (watts) | Time Used per Year | kwh of Electricity | Cost of Electricity |
|---|---|---|---|

3. Follow your teacher's directions about how to get the information needed in each column. You need to determine the power, time used per year, the kwh of electricity used, and the cost of the electricity for each appliance on your lists, using either Method A or Method B.

# Developing an Energy Savings Plan

## Method A (Using Your Own Numbers)

a. Sometimes you can find the power rating (in watts) by looking for a metal tag or label on the back or bottom of the appliance. Copy the number in watts that appears there.

b. You must estimate the amount of hours this appliance is on *in a year*. (Hint: Start with how many hours you think it is on in *one day*, and multiply by 365 to convert to a year. Alternatively, take the number of hours the appliance is on in *one week*, and multiply it by 52 to convert it to a year.)

c. Next find the kwh, or kilowatt-hours, of electricity for each appliance by using the equation:

**(watts/1,000) x time = kwh of electricity**

d. Find the cost for each amount of electricity. Multiply the kwh by the local cost of electricity.

## Method B (Using the Table)

a. Refer to the table, "The Power and Energy Use of Electric Household Appliances." Find each appliance from your top ten list. If an appliance is not listed, try to find something listed that is similar to your appliance. If that is not possible, try to get the information using Method A.

b. Copy the numbers—Power (watts), Time Used per Year, and kwh per Year—that appear for each appliance into the proper columns in your Journal.

c. Find the cost of the amount of electricity that each appliance uses. Multiply the kwh by the local cost for electricity where you live.

d. Now, add together the cost of electricity for each appliance to determine the total cost of electricity used by all of the appliances on each of your lists. Write this number at the bottom of the lists.

*Questions for both methods on page 36→*

## Developing an Energy Savings Plan

# The Power and Energy Use of Electric Household Appliances

**Kitchen**

| Appliance | Power (watts) | Time Used per Year | kwh per Year |
|---|---|---|---|
| Blender | 390 watts | 40 hours | 15.6 kwh |
| Coffee maker | 900 | 120 | 108 |
| Waffle iron | 1,100 | 20 | 22 |
| Toaster | 1,200 | 35 | 42 |
| Broiler | 1,400 | 70 | 98 |
| Range | 12,200 | 100 | 1,220 |
| Microwave | 1,450 | 130 | 189 |
| Dishwasher | 1,200 | 300 | 360 |
| Refrigerator (12 cu ft) | 240 | 3,000 | 720 |
| Frostless refrigerator (12 cu ft) | 320 | 3,813 | 1,220 |
| Freezer (15 cu ft) | 340 | 3,500 | 1,190 |
| Frostless freezer (15 cu ft) | 440 | 4,000 | 1,760 |

**Laundry**

| Appliance | Power (watts) | Time Used per Year | kwh per Year |
|---|---|---|---|
| Iron | 1,000 watts | 140 hours | 140 kwh |
| Washing machine | 500 | 200 | 100 |
| Clothes dryer | 4,800 | 200 | 960 |
| Water heater | 2,500 | 1,600 | 4,000 |
| Quick-recovery water heater | 4,500 | 1,000 | 4,500 |

Activity 51

# Developing an Energy Savings Plan

**Comfort**

| Appliance | Power (watts) | Time Used per Year | kwh per Year |
|---|---|---|---|
| Electric blanket | 180 watts | 828 hours | 149 kwh |
| Air conditioner | 900 | 1,000 | 900 |
| Dehumidifier | 250 | 1,500 | 375 |
| Humidifier | 180 | 900 | 162 |
| Fan (attic) | 370 | 800 | 296 |
| Fan (window) | 200 | 850 | 170 |

**Health and Beauty**

| Appliance | Power (watts) | Time Used per Year | kwh per Year |
|---|---|---|---|
| Toothbrush | 7 watts | 60 hours | 0.42 kwh |
| Hair dryer | 750 | 50 | 37.5 |
| Shaver | 14 | 80 | 1.12 |
| Sun lamp | 280 | 60 | 16.8 |

**Entertainment**

| Appliance | Power (watts) | Time Used per Year | kwh per Year |
|---|---|---|---|
| Radio | 70 watts | 1,200 hours | 84 kwh |
| TV (B/W, solid-state) | 55 | 2,200 | 121 |
| TV (color, solid-state) | 200 | 2,200 | 440 |
| VCR | 120 | 333 | 40 |
| Spa/hot tub | 1,940 | 1,460 | 2,750 |
| Pool pump | 1,380 | 1,848 | 2,830 |

**Housewares**

| Appliance | Power (watts) | Time Used per Year | kwh per Year |
|---|---|---|---|
| Clock | 2 watts | 8,760 hours | 17.5 kwh |
| Vacuum cleaner | 630 | 75 | 47 |
| Sewing machine | 75 | 140 | 10.5 |

**Lighting**

| Appliance | Power (watts) | Time Used per Year | kwh per Year |
|---|---|---|---|
| Light bulbs (on, in home) | 660 watts | 1,515 hours | 1,000 kwh |

## 51

*Developing an Energy Savings Plan*

Meters like this one measure electrical power use.

### Questions

1. Which appliances on your lists cost the most money to operate per year? Which cost the least amount of money?

2. Which list, personal or family, shows the greatest costs? Explain the differences between the lists.

3. Follow your teacher's directions to develop an energy savings plan to reduce 20% of the total cost of electricity used by appliances on a list. (Hint: You need to first calculate how much money you need to save. Then look at the individual appliances on the list and think about reducing the use of each.)

4. How much money will your energy plan save?

5. Explain the choices you made for your energy savings plan. Be sure to describe the evidence you used to make your choices as well as the tradeoffs or conflicts you resolved.

# Activity 52
## Ice and Energy Transfer

Ice-cutting, early 1900s

Sign used to tell the ice man how much ice to deliver

Iceboxes, ca 1870

# 52 Ice and Energy Transfer

## Introduction

### The Ice-Melting Race

This activity will help you understand some basic principles about the energy transfers that occur when we produce and use energy. Think about how the electrical energy used by appliances gets to your home or school. It has to be generated in a power plant, transferred to your home, and transformed into a useful form of energy, such as heat or light. In this activity you will investigate the transfer of heat energy.

## Challenge

You will have 5 minutes to melt as much ice as possible from an ice cube.

## Materials

*For each group of four students:*

- One 10-mL graduated cylinder
- One 50-mL graduated cylinder
- One plastic cup

*For each student:*

- One ice cube
- One plastic bag

*Ice and Energy Transfer*

## Procedure

**Safety**

Do not put the bag in your mouth or under your clothes.

1. Get a plastic bag containing an ice cube. *After you receive the bag, do not leave your seat. Do not remove the ice cube from the bag. Failure to follow these instructions will disqualify you from the race. Handle the plastic bag carefully. Breaking the bag also disqualifies you.*

2. Try to melt the ice in 5 minutes. Your teacher will tell you when to start. Remember: Stay in your seat during the race.

3. When your teacher tells you that your time is up, carefully remove the unmelted ice from the bag and discard it in the cup provided. Carefully pour the water in your plastic bag into the appropriately-sized graduated cylinder and determine how much ice melted. Then discard the melted water so that another student can use the graduated cylinder.

### Questions

1. How many mL of ice were you able to melt in 5 minutes?

2. What energy sources did you use to melt your ice cube?

3. What did you do to increase the rate at which your ice melted? Did other students use different techniques? Compare the rate at which your ice cube melted with the melting rates of other students' cubes, and explain why they may have been different. Be sure you use your knowledge of energy and experimental design in your answer.

4. Describe all of the variables that affect how fast the ice melts. Choose one of these variables only and describe an experiment that will give a fair test of how changing this variable affects the ice melting.

# 52 Ice and Energy Transfer

## Introduction

### The Ice-Preserving Contest

In the ice-melting race, you tried to maximize the transfer of heat to melt the ice. In this activity you will do the opposite: try to prevent the flow of heat to the ice so that the ice will not melt.

## Challenge

Work with a partner to preserve an ice cube for as long as possible.

## Materials

*For each group of four students:*

- One 10-mL graduated cylinder
- One 50-mL graduated cylinder
- One plastic cup

*For each group of two students:*

- One ice cube
- One plastic bag
- One empty half-gallon milk carton
- Materials from home for preserving your ice cube

Activity 52

# Ice and Energy Transfer

## Procedure

1. Set up your ice-preserving experiment in the container provided and allow it to remain undisturbed until your teacher tells you that time is up. *You will be disqualified if your plastic bag with the ice cube leaks.*

2. Carefully pour the water melted from the ice into the appropriately-sized graduated cylinder. Record the volume of water you obtained.

## Data Processing

1. Describe what you did to preserve your ice. Explain why you thought it would work.

2. How much water resulted by the end of the class period?

3. What would you do differently in another ice-preserving contest?

# 52 Ice and Energy Transfer

## Introduction

### Refrigerators, Iceboxes and Heat Transfer

Keeping our homes comfortable and our food from spoiling involves the transfer of heat. In some cases we want to encourage heat transfer, while in other cases we want to prevent it.

## Challenge

Think of the old-fashioned icebox in terms of desirable and undesirable transfers of heat as you answer the questions that follow the reading.

It's the hottest day of the summer. Your friend comes to visit, and the two of you decide to go outside and relax. On your way outside, you decide to stop in the kitchen for some cool drinks. You open the refrigerator and find some ice-cold sodas. You take the sodas outside and alternate between sipping your cool drink and holding it against your hot forehead. Some of the heat from your body transfers to the cool drink. It feels great.

Living in the 20th century has some distinct advantages, including the availability of modern refrigerators. The refrigerator is America's most common appliance. It is found in more than 99.5% of all homes in the United States. Refrigeration helps us to enjoy cool drinks on hot days. More importantly, it helps prevent food from spoiling.

# Ice and Energy Transfer

Food spoilage has always been a health risk. In the past, food poisoning resulted in many more cases of illness and death than today. However, even today, public health officials believe that millions of people suffer from food poisoning in the United States each year. While most people recover in a day or two, some people, especially children, the elderly, or those weakened by other illnesses, die as a result of food poisoning. Refrigeration and freezing are two of the best ways to prevent food from spoiling. Only refrigeration maintains the taste of fresh food. Other methods change the flavor and texture of the food.

Long ago snow and ice, cool streams, springs, caves, and cellars were used to refrigerate food. The Chinese cut and stored ice in 1000 B.C. About A.D. 1300, Marco Polo described ices and sherbets he had eaten in the Far East. The idea of "manufacturing" ice dates back to Venice in the 16th century, when it was discovered that adding salt to ice produced a slushy brine that remained below the freezing temperature. This could then be used to freeze clear water into solid ice. This is also the old-fashioned way of making ice cream.

In the United States, using ice to preserve food in homes became popular in the mid-1800s. At first, the ice trade provided "natural" ice from northern lakes and rivers to cities in the south. Eventually wooden ice boxes lined with tin or zinc and insulated with various materials including cork, sawdust, and seaweed were used to hold blocks of ice to "refrigerate" food. A drip pan collected the meltwater—and had to be emptied daily.

A good icebox prevented transfer of heat from the surrounding room while allowing for good transfer of heat from the food to the ice. One problem with iceboxes was that the ice had to be replaced every few days.

Warm winters in 1889 and 1890 resulted in severe shortages of natural ice in the United States. This increased the use of mechanical refrigeration for the dairy and meat packing industries and for the freezing and storage of fish. Commercial refrigeration

## Ice and Energy Transfer

techniques were also applied to railroad cars and were used in "coolers" in grocery stores as well as in various ways in the textile, paper, drug, soap, liquid gas, sugar, and munitions industries.

Mechanical refrigerators for the home began to appear on the U.S. market between 1910 and 1915. By 1930, half of the homes in the United States had refrigerators. This number increased to 85% by 1944. It's hard to imagine life without one, isn't it?

### Questions

1. Draw a picture of a room with an icebox containing a block of ice. Think of the room, the icebox, the ice, and the food as a system. The icebox is a smaller system within the larger system of the room. On your diagram, use arrows to show all of the heat transfers possible between parts of the large system. Use a solid line to represent transfers that help to keep the food cold. Keeping in mind that the purpose of the icebox was to keep food cold, indicate for each arrow whether the heat transfer it shows contributed to that purpose. Explain why.

2. In order to make the ice in an icebox last longer, some people would wrap a blanket or newspapers around the ice to keep it from melting. Keeping in mind that the purpose of the icebox is cooling food, explain why this was or was not a good idea. Use a diagram like the one you used for Question 1 to explain your answer.

Activity 52

# Activity 53

## Quantifying Energy: Calorimetry

### Introduction

**Mixing Hot and Cold Water**

In Activity 52 you learned about heat transfer from a hot object, your hand, to a cold object, ice. Now you will learn one way to measure the amount of heat transferred.

### Challenge

Measure the amount of heat lost by hot water and the amount of heat gained by cool water. Compare the amounts.

### Materials

*For each group of two students:*

- One 50-mL graduated container
- One foam cup
- One metal-backed thermometer
- Source of hot and cool water
- One plastic spoon

# Quantifying Energy: Calorimetry

## Procedure

1. Before conducting the experiment pictured below, be sure that you have described in your Journal what you think will happen to the temperature of the water when you mix the hot and cool water. Then prepare a data table in your Journal, like the one shown here, to organize your observations of what actually happens.

| Volume | | Starting Temperature | | Final Temp. | Temp. Rise | Temp. Drop |
|---|---|---|---|---|---|---|
| Cool Water | Hot Water | Cool Water | Hot Water | Mixture | Cool Water | Hot Water |
| 60 mL | 60 mL | | | | | |
| 60 mL | 30 mL | | | | | |

Follow your teacher's instructions for obtaining 60 mL of cool water and hot water in each of your foam cups.

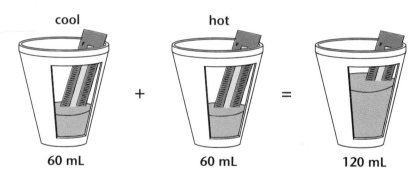

2. Take the temperature of the cool water. Wait for a few seconds until the temperature remains steady and then record it in the data table.

3. Take the temperature of the hot water and record it. *Quickly* add the cool water to the hot water. Stir gently with a spoon until the temperature of the mixture remains steady. Record the temperature in your data table.

# Quantifying Energy: Calorimetry

4. Calculate the temperature change for both the hot water and the cool water. Fill in the appropriate columns in the table.

5. Now do a similar experiment but cut the amount of hot water in half, to 30 mL. Describe in your Journal what you think will happen before testing it. The results from the previous experiment should increase your ability to predict the outcome.

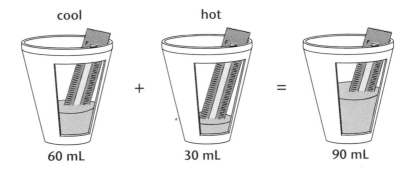

 **Data Processing**

1. When you mixed equal volumes of hot and cool water, what happened to the temperature? How did the temperature rise of the cool water and temperature drop of the hot water compare?

2. Was the result what you expected? What does this tell you about the energy transfer in this activity?

3. When you mixed only 30 mL of hot water with 60 mL of cool water, how did the temperature rise of the cool water and temperature drop of the hot water compare? How would you explain your results?

# 53

## Quantifying Energy: Calorimetry

### Introduction

**Measuring the Heat Used to Melt Ice**

Sometimes heat energy can be converted to another form of energy. When water is used to melt an ice cube, the heat energy of the water is transferred to the ice. The amount of energy used to melt the ice can be calculated by measuring the temperature change of the water and calculating the heat lost by the water as you did in the previous investigation.

### Challenge

Measure the amount of heat needed to melt an ice cube.

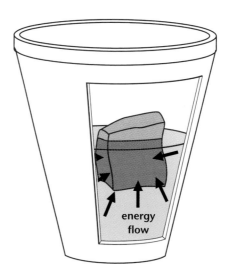

# Quantifying Energy: Calorimetry

## Materials

For each group of two students:

- One 50-mL graduated cylinder
- One foam cup
- One ice cube in a 9-ounce plastic cup
- One metal-backed thermometer
- Source of cool water
- One plastic spoon

## Procedure

1. Follow your teacher's directions for obtaining an ice cube and some hot water.

2. Measure 100 mL of hot water into the foam cup.

3. Record the temperature of the hot water. Then immediately shake any water off of the ice cube and place the ice in the hot water.

4. *Gently* stir until the ice is completely melted. (This can be difficult to see, so look carefully!) *As soon as the ice has melted,* record the temperature of the water.

# Quantifying Energy: Calorimetry

**Data Processing**

Remember, the energy required to melt the ice cube came from the heat energy of the water surrounding it. The amount of heat lost by the water is equal to:

**mass of water (grams) x temperature change of water (°C)**

1. What was the temperature change of the water?

2. How many grams of hot water were cooled? (Hint: You used 100 mL of hot water.)

3. Calculate the heat gained by the ice as it melted by multiplying the temperature change of the water (your answer for Question 1) times the mass of the water (your answer to Question 2).

4. Your measurement of the heat needed to melt the ice is based on the assumption that the heat gained by the ice is exactly equal to the heat lost by the water. Do you think all of the heat lost by the hot water was transferred to the ice? Explain. You may use a diagram similar to the one on page 48 as part of your explanation.

5. Based on your answer to Question 4, do you think you are more likely to have overestimated or underestimated the amount of heat needed to melt an ice cube?

6. What effect did stirring have on the melting of the ice? What would happen if you did the same experiment again but didn't stir the ice?

*Quantifying Energy: Calorimetry*

# 53

## Peanut Calories

### Introduction

A peanut contains stored energy. That energy is released when we eat and metabolize the peanut. The amount of energy we can obtain by eating the peanut is usually measured in food Calories. A food Calorie (with a capital "C") is equal to 1,000 calories.

### Challenge

Determine the number of calories in one peanut. Then convert the calorie measurement into food Calories.

### Materials

*For each group of four students:*

- One wire coat hanger, wrapped with aluminum foil
- One aluminum beverage can
- One piece of clay
- One straight pin
- One peanut, removed from the shell
- One 50-mL graduated container
- One glass thermometer
- Water
- Matches

Activity 53

## Quantifying Energy: Calorimetry

### Procedure

If you burn a peanut, it will give off energy in the form of heat and light. The heat can be measured by determining how much it raises the temperature of a known volume of water. Remember, the heat absorbed by water is equal to the mass of the water (in grams) times the temperature change of the water (in 8 °C).

 **Safety**

Be sure to use safety eyewear during this investigation. Have a cup of water available in case you have to extinguish the peanut flame. The can may become quite hot. Carefully follow all instructions from your teacher. Be especially careful not to get long hair or clothing near the flame.

1. Carefully insert the pin through the peanut. (If the peanut falls apart, get another one.)

2. Place the pin in the clay.

3. Set up the calorimeter with 100 mL water in the can.

4. Just before burning the peanut, record the starting temperature of the water.

5. Light the peanut. Once it begins to burn, slide it under the can and let it burn completely.

6. As soon as it stops burning, stir the water and record its final temperature.

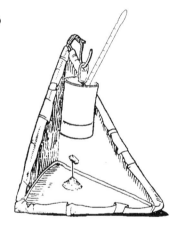

# Quantifying Energy: Calorimetry

 **Data Processing**

Before calculating the calories of heat given off by the peanut, you need to know (and record in your Journal):

1. The amount of water used = _____ mL = _____ grams

    *Hint: One mL of water weighs one gram.*

2. The temperature change of the water = _____ °C.

3. Now you can calculate the calories by multiplying your answers for Items 1 and 2 above:

    **calories = grams water X temperature change of water**

    Show your calculation in your Journal.

4. Your calorie determination is only as good as the calorimeter you used. Explain why.

5. Based on what you have learned about measuring heat, how would you improve the design of the calorimeter so that you could measure all of the heat released by a burning peanut? Draw a detailed diagram of your improved calorimeter, and explain why it would be better than the old one at measuring the energy released by the burning peanut. Be sure to design a calorimeter that you could build yourself, if you had the materials. Label the parts of your calorimeter on the diagram.

6. Do you think your calculated peanut calorie figure is likely to be lower or higher than the actual value? Explain.

7. A food Calorie (notice the capital "C") is equal to 1,000 calories, or one kilocalorie. How many food Calories are there in one peanut?

## Extension

With your teacher's permission, determine the number of calories in a marshmallow or a puffed cheese snack.

*Activity 53*

# Quantifying Energy: Calorimetry

## Introduction

### Energy Gathering and Farming

Gathering energy is something that plants and animals do in order to survive. Plants' energy comes from the Sun. Like plants, some animals don't have to go out and gather energy, it comes to them. For example, clams filter bits of organic matter that float by in the water. But, most animals do go out and search, which takes quite a bit of energy itself. For over a million years our human ancestors gathered energy in this way. Even today, many people still hunt, fish, and/or gather, especially in poor countries. But today, in some places, the energy sources are so limited that there is not enough to go around, even after traveling miles in the search for food.

## Challenge

Reflect on how people's methods of obtaining food have changed. Think about how this relates to our changing use of energy and its impact on the Earth.

*Food Energy Ratios*

| Method of Obtaining Food | Ratio of Energy Input to Energy Output |
|---|---|
| Hunting and gathering | 0.1 |
| Low-intensity corn | 0.2 |
| Range-fed beef | 0.5 |
| Intensive corn | 0.5 |
| Coastal fishery | 1.0 |
| Feedlot beef | 10.0 |
| Distant fishery | 14.0 |

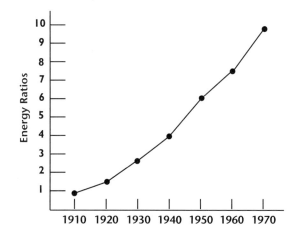

*U.S. Food Energy Ratios 1910–1970*

Based on data in J.S. Steinhart and C.E. Steinhart, "Energy Use in the U.S. Food System," Science, April 19, 1974, pp. 307–315.

# Quantifying Energy: Calorimetry

Unlike hunting and gathering, farming is a more recent human development. Researchers believe that the origin of agriculture was about 10,000 years ago. Farming allows great numbers of people to stay in one place and do other things during the day besides hunt and gather food (like go to school). But the job of farming, like all work that is done, requires energy. Pick a type of farming (such as citrus farming in Florida, dairy farming in Wisconsin, or peanut farming in Georgia) and think about all the ways that energy is used to produce and deliver the product to market.

What would you say if the energy it took to gather or farm a particular food was greater than the energy contained in the food itself? The chart on the opposite page shows the ratio of energy input to food energy output for several foods produced in different ways. A ratio of 1.0 means that the input and output energy is the same. For example, coastal fishing has a ratio of 1.0—the input and output energy are about the same. But distant fishing requires much greater energy input—consider all of the ways that energy is used to get to the fishing grounds, catch the fish, process the fish, store the fish for the return, and then get the fish to market. Many modern methods of obtaining food, such as distant fishing and intensive agriculture, have high energy input to energy output ratios.

## Questions

1. In general, what types of food cost more energy to produce than they provide? (Hint: Look at the foods above the food-energy ratio of 1.0.)

2. Can you think of reasons why the foods above the ratio of 1.0 take more energy to produce than they release as chemical energy in the body?

3. With the passage of time, distant fishing has become more common than coastal fishing. Why? Why is the energy input so much greater for the distant fishing?

4. Notice the changing food-energy ratio for the U.S. food system as a whole from 1910 to 1970. How has it changed? What changes in society might explain this change?

# Activity 54

# Efficiency: Energy Changes and Waste

## Introduction

### Hot Bulbs

In this activity you will calculate how efficient a flashlight bulb is at producing light energy. The more efficient the bulb is at producing light energy, the less waste heat it produces. Be sure to record and label all measurements and calculations in your Journal. A chart similar to the one on page 58 will help you keep organized.

## Challenge

How good is a flashlight bulb as a heater? Do the activity to find out!

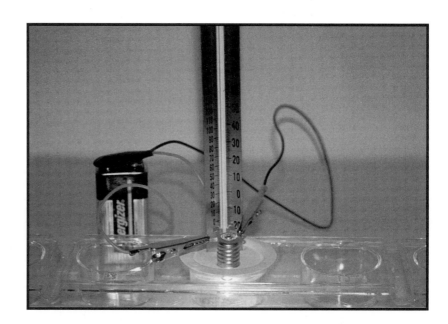

This apparatus is used for determining the energy efficiency of a flashlight bulb.

# Efficiency: Energy Changes and Waste

## Materials

*For each group of four students:*

- One 9-volt battery, with harness and connection clips
- One flashlight bulb with socket and white plastic cover
- One SEPUP tray
- One graduated cylinder
- One thermometer

## Procedure

### Safety

Do not try this investigation with any other kind of battery without consulting your teacher. **Never, under any circumstances, place plugged-in electrical appliances in or near water.**

1. Using the graduated cylinder, carefully measure 12 mL of water into Cup A of the SEPUP tray. Measure the initial temperature of the water and record it in your Journal.

2. Use Cup B as a control for this experiment. Decide in your group what should be placed in Cup B and what measurements should be taken for the control.

3. Insert the flashlight bulb into the white plastic cover. Be sure that the concave side of the plastic cover faces up and the bulb faces down. Screw the brass socket onto the bulb. Insert the thermometer into the plastic cover as shown on the opposite page.

4. Connect the 9-volt battery to the socket, using the connectors provided. Place the lighted bulb into the water in Cup A for exactly 3 minutes. Time this as precisely as you can.

5. After 3 minutes, remove the bulb from the cup. Measure the final temperature of the water and record it.

# Efficiency: Energy Changes and Waste

**Data Processing**

1. Calculate the temperature change of the water (final temperature minus the initial temperature).

2. Calculate the heat energy output of the flashlight bulb (in calories) using the equation:

**heat energy output (in calories) = volume of water x temperature change**

3. Using the given energy input of 82 calories in 3 minutes, calculate the waste heat production of the flashlight bulb using the equation:

**% waste heat = (heat energy output / energy input) x 100**

Record and label the calculation in your Journal.

4. Now use the heat efficiency calculation you just made to state the light bulb's light efficiency.

*Heat Efficiency of a Flashlight Bulb*

| Measurements and Calculations | Experiment | Control |
|---|---|---|
| | Cup A | Cup B |
| Volume of water (mL) | 12 mL | |
| Initial temperature (°C) | | |
| Final temperature (°C) | | |
| Temperature change (°C) | | |
| Time bulb was on | 3 minutes | 0 minutes |
| Heat output of bulb | | |
| Energy input to bulb | 82 calories | |
| % waste heat produced by bulb | | |
| % light efficiency of bulb | | |

*Efficiency: Energy Changes and Waste*

# 54

 **Questions**

1. What was the % waste heat produced by your flashlight bulb? From your results, how efficient are these bulbs at producing heat? At producing light?

2. Why should you use a control cup (in this experiment Cup B)? What did you place in the control cup, and what measurements did you take? Explain.

3. A typical light bulb is nearly 90% efficient at producing heat. Does your answer agree with the 90% figure? Why not? What problems did you have? What would you do differently?

Activity 54

# Efficiency: Energy Changes and Waste

## "You can't get something for nothing..."*

The efficiency of a light bulb is shown in the pie chart below. The efficiencies of other devices are given in the table.

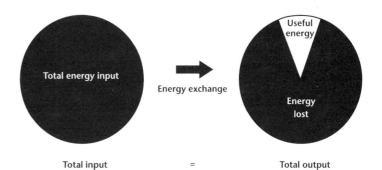

Total input = Total output

| Device | % Efficiency | % Waste Heat |
|---|---|---|
| Fluorescent light (20-watt, 24-inch) | 50 | 50 |
| Incandescent light (40-watt) | 11 | 89 |
| Auto engine (gas) | 30 | 70 |
| Auto engine (diesel) | 35 | 65 |
| Coal power plant | 38 | 62 |
| Nuclear power plant | 31 | 69 |
| High-efficiency gas furnace | 90 | 10 |
| Typical gas furnace | 75 | 25 |
| Typical oil furnace | 63 | 37 |

Use data from the table to create additional pie charts for the energy efficiency of other devices.

*...and you can't break even either."

Activity 54

*Efficiency: Energy Changes and Waste*

# 54

## Introduction — A Cool Energy Decision

Everyone uses a lot of light bulbs, but few people think much about them. When one bulb burns out, you replace it. And that's probably the last you think about light bulbs until the next time one burns out. This reading is about an important energy decision: replacing light bulbs.

## Challenge

Which kind of bulb should you use to replace a burned out light bulb? Incandescent or compact fluorescent? Read on and then decide.

Which bulb is more economical? The incandescent (left) or the compact fluorescent (right)?

Activity 54

# Efficiency: Energy Changes and Waste

The light bulb most often used in our homes is called an incandescent bulb. In this activity we discovered one of its problems: it is very inefficient. The typical light bulb produces as much as 90% heat energy and only 10% light. That means the bulb uses a lot of electricity to heat the room rather than to provide light. A second problem is related to all the heat energy produced by the bulb. The bulb produces light by heating up a filament inside. This filament is actually burning, and when it burns up and breaks, the light bulb goes out! Because the bulb produces light by heating the filament so that it glows very brightly, incandescent light bulbs don't last very long.

Compact fluorescent lamps, however, produce light in a different way. The electricity adds energy to mercury vapor inside the tube. (Scientists call this *exciting the gas*.) The mercury vapor then releases energy, which hits a special coating called *phosphors*. This phosphor coating is spread all along the inside of the tube and converts the energy to visible light. So there is no filament, and no burning. This type of light bulb creates a lot less heat energy.

## Questions

Note: Your teacher will help you complete a table in your Journal like the one shown here that compares incandescent and compact fluorescent light bulbs. Answer each of the questions based on the information in the table. Show your work as evidence for your decision.

*Comparison of Incandescent and Compact Fluorescent Lights*

| Type of Light Bulb | Cost to Buy | Power Rating | Average Lifetime |
|---|---|---|---|
| Incandescent |  | 75 watts | 750 hours |
| Compact fluorescent |  | 18 watts | 7,500 hours |

# Efficiency: Energy Changes and Waste

*If compact fluorescents are so great, why don't more people use them? Let's find out more about this energy decision.*

1. Compare the cost to buy a 75-watt incandescent light bulb and an 18-watt compact fluorescent bulb. (Both bulbs provide about the same amount of light.) Based on this information, what decision would you make when replacing a burned out 75-watt light bulb? Would you buy a compact fluorescent light bulb or an incandescent one? Explain your choice.

2. Now compare the average lifetime for each type of light bulb. One compact fluorescent bulb lasts as long as how many incandescent bulbs? Using this information, what energy decision would you now make? (Hint: Find out the *total* cost of these incandescent bulbs compared to the cost of one compact fluorescent.) Explain your decision.

3. Now let's look at another factor. Let's compare the amount of electricity both bulbs use in a specific amount of time. Use 7,500 hours (the lifetime of one compact fluorescent bulb) as the specified time.

   a. How many watt-hours of electricity would one compact fluorescent bulb use over its lifetime of 7,500 hours?

   b. How many watt-hours of electricity would an equivalent number of incandescent light bulbs use over 7,500 hours?

   To calculate watt-hours of electricity use the equation:

   **watts x time = watt-hours of electricity**

   c. Electricity is usually charged by the kilowatt-hour of use. Change your results in watt-hours to kilowatt-hours by dividing by 1,000.

*More questions on other side of this page →*

## Efficiency: Energy Changes and Waste

**Questions**
*(continued)*

4. Your teacher will tell you the cost of electricity in your area. Write it in your Journal.

    a. What will be the total cost of the electricity for the incandescent light bulbs?

    b. What will be the total cost of the electricity for the compact fluorescent bulb?

    (Hint: Use the kilowatt-hours of electricity you found for each type of bulb and multiply it by the cost of electricity.)

5. Find the *total* expense for using each type of bulb. Add together the cost of the bulb(s) and the cost of the electricity. Using all this information, which bulb is a better buy? Use your calculations to explain your decision.

6. What additional information would you like to know before making a final decision about which bulb to buy?

7. Which bulb do you think most people buy when they go into a store? Why? If you were a manufacturer of compact fluorescent bulbs, what would you do to get people to buy your bulbs?

# Activity 55
# Electrical Energy: Sources and Transmission

## Introduction

### Sources

There are many ways to generate electricity. Some use renewable resources such as biomass while others use nonrenewable resources such as coal, natural gas, and fuel oils.

## Challenge

Suggest ways to generate electricity and explore some advantages and disadvantages for each energy source.

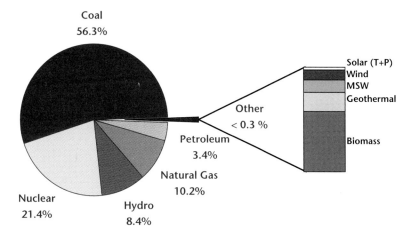

**U.S. Electrical Energy Sources 1994**

- Coal 56.3%
- Nuclear 21.4%
- Hydro 8.4%
- Natural Gas 10.2%
- Petroleum 3.4%
- Other < 0.3% (Solar (T+P), Wind, MSW, Geothermal, Biomass)

| | |
|---|---|
| **Geothermal energy:** | The underground heat of the Earth furnishes energy. |
| **Biomass:** | Wood, paper, or municipal wastes produce energy. |
| **Solar thermal (T) energy:** | Mirrors and reflectors concentrate the sun's heat on water, oil, or other materials to generate energy. |
| **Solar photovoltaic (P) energy:** | Solar cells (batteries) convert the energy directly into electricity. |
| **Municipal solid waste (MSW):** | The incineration of municipal solid waste produces energy. |
| **Hydro:** | Flowing or falling water generates power. |

# Electrical Energy: Sources and Transmission

### Wind Energy

Wind can turn the blades of large turbines to produce energy. Wind farms are set up in areas of the country where the wind blows for long periods of time. From 1880 to 1930, over 6 million windmills generated electric power in rural areas of the western United States before rural electrification. Today, vast areas of land must be developed for wind turbines to produce significant amounts of electric power.

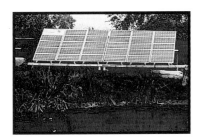

### Solar Energy

Energy from the sun can be directly converted into electricity by using solar cells (photovoltaics). It can also be used to heat water (thermal energy) and other liquids, causing them to transfer their heat energy to other sources, such as swimming pools, hot-water heaters, and turbines to produce electricity. The production of solar cells creates hazardous wastes. For large scale applications, solar collectors require large areas of land.

### Hydroelectric Power

Hydroelectric power generators use water to spin a turbine connected to a generator. The water may come from rivers, streams, or lakes. Construction of a dam is often necessary. The kinetic energy in the flowing or falling water is converted to mechanical energy by the turbine and then into electricity by the generator. Some people are concerned about the environmental effects of blocking streams, rivers, or lakes to create dams.

### Biomass Energy

Biomass refers to the production of power by burning plants or waste material. Power plants that use biomass (wood, paper, or municipal waste) are similar to power plants that burn coal, oil, and natural gas. The fuels are combusted in a boiler and produce steam to drive a conventional steam turbine. The production of biomass energy has some of the same environmental concerns as incinerators.

# Electrical Energy: Sources and Transmission

### Fossil Fuels (Coal, Fuel Oil, and Natural Gas)

Coal is the most abundant fossil fuel in the United States and the leading source of electricity. Coal is powdered into a fine dust and injected into the furnace to power a conventional steam boiler. Fuel oil and natural gas are also used as fuels to boil water. The gases produced by the combustion of coal—sulfur dioxide and nitrogen oxides—have been the major producers of acid rain. Natural gas produces only water and carbon dioxide when burned "cleanly." Fossil fuels are nonrenewable energy sources.

### Geothermal Energy

Energy from below the surface of the Earth, created by the heating of rocks or mud, is used to heat water to produce steam which drives conventional turbines.

### Nuclear Power

When atoms of uranium are split apart (fission reactions), the heat generated can be used to boil water and produce steam to power a turbine. Concern over environmental effects and accidents, such as the accidents at Chernobyl in the former Soviet Union (1986) and Three Mile Island in Pennsylvania (1979), have brought construction of new plants in the United States to a standstill. Nuclear power plants do not produce the air pollution of fossil fuel burning plants. They do generate radioactive waste that requires special facilities for disposal.

### Question

Your city needs more electric power to provide new jobs and more efficient services. You have been asked to recommend a source of power. Considering all of the different sources, which one would you choose and why? Be sure to identify the various sources of electrical power available as well as the factors you would consider in making your choice.

# Electrical Energy: Sources and Transmission

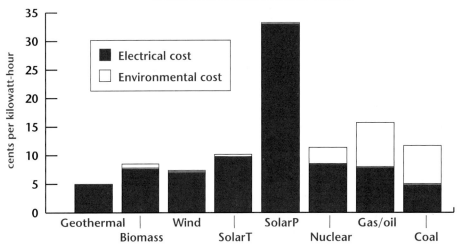

Source: American Council for an Energy Efficient Economy, 1990

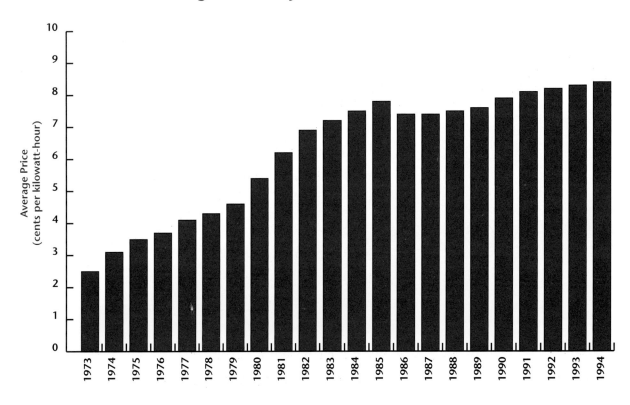

# Activity 56

## Energy From the Sun

### Introduction

**Exploring Solar Energy**

The sun sends a constant amount of energy to the Earth's surface. How can we transform this energy to heat water and to provide electrical power?

### Challenge

Construct a solar water-heating system and determine its efficiency. Then prepare a group report describing your recommendations for the redesign of the collector for maximum efficiency.

### Materials

*For each group of four students:*
- One miniature water pump
- Black plastic tubing
- One white plastic tubing holder—coil-shaped
- One alkaline D-cell battery
- One D-cell battery holder with leads
- One set of leads with alligator clips (one red and one black)
- Two metal-backed thermometers
- One 50-mL graduated cylinder
- Two low-profile foam cups
- One 9-ounce plastic cup
- Black electrical tape
- Masking tape
- One graph paper transparency
- One non-permanent transparency pen

# Energy From the Sun

## Procedure

**Safety**

Be careful not to break the thermometers. The metal on the thermometers may become quite hot in the sun.

1. Use the length of black tubing to construct a coil that fits within the coil impression in the plastic holder. Use black electrical tape to hold the coil together and to prevent it from moving out of the channels in the plastic holder. *Be sure to leave 25 cm (10 inches) of tubing on each end free of the coil.* The pump will transport the heated water to and from the cup through the ends of the tubing. Refer to the diagram below as a guide.

2. You may want to adjust the angle of the plastic holder towards the sun. You can use a book or two to prop up one side of the holder. You may also wish to tape your collector to a piece of cardboard to give it more support.

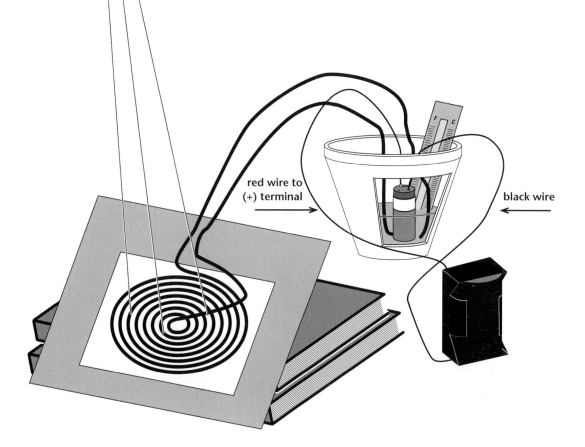

## Energy From the Sun

3. Attach one end of the black tubing to the pump by pushing it into the cylindrical opening on the side of the pump's base.

4. Place the pump, upright, in one of the foam cups. Avoid pinching the tubing.

5. Place the other end of the black tubing in the cup.

6. Pour 100 mL of water into another foam cup. This is your control. Record the temperature of the control.

7. Place the battery in the holder so that its tip (positive end) faces the side with the red connector. Attach the red lead to the red wire, and then connect the red lead to the positive (+) terminal of the pump (look for the small plus sign). Now attach the black lead to the black wire, *but don't attach it to the pump yet!*

8. Measure 100 mL of water into the plastic cup. Carefully pour just enough water from the plastic cup into the foam cup containing the pump to cover the base of the pump. Save the rest to use in Step 10.

9. Take the temperature of the water in the cup with the pump. Leave the thermometer inside the cup. Do the same with your control cup. Leave both of the thermometers inside the cups.

10. To start the water pump, connect the black lead to the pump and begin your timing. As water goes from the cup into the tubing, slowly add the rest of your 100 mL of water. *Do not allow the water level in the cup to rise more than 1 cm above the base of the pump.*

11. Allow the pump to run for 15 minutes. Record the temperature of the water in both cups every minute. After 15 minutes, disconnect the leads from the pump.

12. Record the final temperature of the water in both the control cup and the experimental cup. Clean up as directed by your teacher.

# Energy From the Sun

 **Data Processing**

1. Make a graph of the temperature changes over time for both the control and the experimental cups.

2. If you used 100 mL of water in the cup, calculate how many calories of energy were absorbed from the sun by the tubing and were transferred to the water.

3. Calculate the surface area of the tubing holder, which is approximately 16.5 cm by 19.0 cm.

4. Do you think you should use the whole surface area of the collector in your calculations? Why or why not?

5. If the sun supplies 1.5 calories of energy per square centimeter per minute to the Earth, how many calories were supplied to your collector for 15 minutes?

6. Now calculate the efficiency of the solar heating system. The efficiency would be the number of calories actually gained by the water in the system divided by the total number of calories supplied to the system by the sun.

7. As a group, write a short report describing the outcomes of this experiment. Include the graph and calculations you did above. Also include suggestions on how you would redesign your collector to increase its efficiency.

*Energy From the Sun*

# 56

## Introduction

### Solar Energy

The sun is the ultimate source of the energy we use. It is a renewable source of energy. In this reading you will learn how this abundant source of energy can be used to heat swimming pools.

## Challenge

> Use the information from this reading to help you understand how to construct a better hot-water collection system using the sun's "free" energy.

The energy costs of heating an outdoor swimming pool with an electric or gas water heater are very high. What's worse, the energy used to heat the water in the pool quickly transfers to the air around it.

Fortunately, outdoor swimming pools are warmed passively by the sun when the sun's radiant energy falls on them. Some of the sun's energy is transferred to the water and heats it while some is reflected off the surface, in much the same way a mirror reflects the image of your face. This warming by the sun is considered passive because you do not have to do anything special to the pool to warm it. Of course, the amount of heating is usually not enough to get most people to dive in, at least not in the morning!

A solar-heating system is used to increase the amount of energy actually transferred to the water to heat it. In a total system, this would also prevent, as much as possible, the loss of that energy to the outside environment. This is usually done by one or a combination of two methods. One method is called a *passive system*. With a passive system, a translucent plastic cover with air cells in it is placed over the pool. This system works well because some light energy can still pass through the cover to heat up the

Activity 56    73

# Energy From the Sun

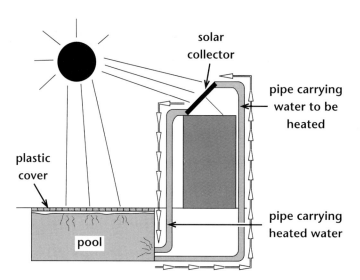

water, but the cover also acts like a heat insulator to slow down the rate of heat loss to the air. Using such a cover at night, or when it gets cool, is especially effective at reducing energy costs.

The second system, called an *active system,* involves placing black plastic panels with water-flow channels in them on the roof of a nearby building in a specially constructed holder. The panels are aimed towards the sun for as much of the day as possible. Water is pumped from the pool up to the panels and flows through the channels. This works well because the black plastic is an excellent absorber of radiant energy, so the water flowing through the panels is warmed and returned to the pool. This is a *closed system* because when the sun is out, all of the water in the pool automatically flows through the panels as part of the pool's water circulation system. Heat sensors are used to automatically turn the system on whenever the water standing in the channels reaches a certain temperature.

Although an active system uses electrical energy to pump the water up to the panels, this is accomplished by the same system that circulates and filters the pool water. A well-designed system will heat the pool more effectively than a passive system made up of just a cover. Combining systems provides both heating from the active system when the sun is shining and prevention of heat loss at night when the pool is covered.

# Energy From the Sun

## Introduction
### The Photovoltaic Effect

You will investigate how a solar cell can directly transform solar energy into electricity.

## Challenge

Consider the tradeoffs that would be involved if all electricity were generated by this method.

Solar photovoltaic panels

Activity 56

*Energy From the Sun*

## Materials

*For each group of four:*

- One electric motor with tape on axle
- One solar cell
- Two wire leads with alligator clips (one red and one black)
- Small piece of cardboard

## Procedure

1. Work together in your group to set up and test the solar electric cell. Use the diagram below as a guide. Describe the various conditions—angle to the sun, amount of sunlight, electrical connections—that produce the fastest spin of the motor. How can you measure the rate of spin?

2. Now join with another group to connect two solar cells together to drive one motor. Your goal is to furnish more energy (voltage) to the motor. As a group, decide how you might go about doing this and try it out. Discuss your results with other groups.

   To simulate the effect of cloud cover, use the piece of cardboard to block off first one-quarter, then one-half, and finally three-quarters of the surface of the two solar cells. Keep a record of what effect this has on the speed of the motor.

Activity 56

# Energy From the Sun

### Data Processing

Summarize in your Journal what you have learned about solar cells. Include a short statement recommending or rejecting solar cells as your sole source of energy. Be sure to give evidence for your reasoning.

### Extension

If your teacher permits it, drive the solar water-heater system from the previous investigation with solar cells rather than the D-cell battery. Will one solar cell power the pump, or does it take several? When you used the battery, you were not able to control the speed of the pump; with the solar cell you can control the speed. Can you use the solar cell to determine at what rate of pumping the water heats more quickly? Should the water circulate quickly or slowly?

# Activity 57: Controlling Radiant Energy Transfer

## Introduction

### A "Reflective" Question

In the last activity, you did investigations to find out how to capture and store energy from the sun. In this activity, you will look at methods for preventing unwanted heating by the sun.

## Challenge

Learn how the transfer of energy from the sun can be controlled by special materials.

## Materials

*For each group of four:*

- Two metal-backed thermometers
- One piece of clear plastic film
- One piece of reflective plastic film
- Two prefolded boxes
- Masking tape (approximately 30 cm or 12 inches)

78  Activity 57

# Controlling Radiant Energy Transfer

## Procedure

Look at your two pieces of plastic film. Plan an investigation that will allow you to compare the ability of these two plastic films to prevent the transfer of the sun's heat through a window into a room. Use the box and thermometer as shown below to simulate a room with a window. Carry out your investigation.

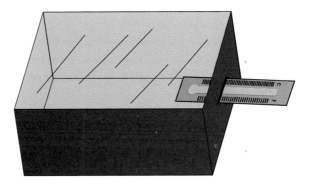

## Data Processing

1. Write a summary describing the results of your investigation.

2. Think about the investigations you did in the last two activities. There were some similarities and some differences between the investigations. Describe how you used materials to accomplish the different energy transfer goals in Activities 56 and 57.

   Be sure to identify:

   a. the energy chains involved;

   b. the effect of the materials chosen on the energy transfer; and

   c. the results of the decisions made in each case.

# Activity 58 — Designing an Energy-Efficient Car

## Introduction
### The Future is Now

Automobiles have changed the way we live. But the products from the combustion of gasoline have also transformed the world.

## Challenge

> Read about Mexico City today and in a possible future. Assuming that cars are a major source of pollution, could this future become a reality someday in large cities across America?

March 9, 2033

John's hovercraft shot smoothly over the black volcanic peaks, and then set a straight course for the deserted city ahead. It had been nearly 15 years since he had last been here, in 2018, just before the massive evacuation. Now the city, with its tall towers and once-green parks, busy roads, and cluttered markets, sat empty against the horizon of brown air. He had a few moments before he had to put on his gas mask. He remembered Ricardo's face when Ricardo and his family had to flee the city. Where was Ricardo now? John looked down at the old journal sitting in his lap and opened it to a page his dad had written in 2016. He began to read...

# Designing an Energy-Efficient Car

> When Ricardo Rivas rides his bicycle to deliver the Sunday paper to a fashionable neighborhood in the city, he wears a surgical mask. His doctor recommended that he wear it because of his chronic cough. And because he wants to be a singer someday, he continues to wear the mask, even though the worst of the winter smog season has passed. He has good reason to be cautious. He remembers the day last winter when he saw birds falling from the sky, poisoned by the air pollution. Each day, toxic substances like carbon monoxide, lead, hydrocarbons, nitrogen oxides, sulfur dioxide, ozone, and soot are released into the atmosphere. The stubborn brown cloud of smog has made headaches, eye irritations, and respiratory problems a foul fact of life for the city's 25 million residents. Over 75% of the air pollution is a result of the over 3 million automobiles, taxis and buses in the city. Many of these vehicles are old, with little or no pollution control.

John put the journal down and thought, "Why did this have to happen? Was there any way that people could have changed their cars or their driving habits to save the city?" He thought again about Ricardo as he climbed out of the hovercraft. The deserted city lay silent, dead.

---

Does the story you just read sound impossible? With new programs aimed at reducing vehicle emissions, it is hoped that a scenario like the one just described can be avoided. However, it will take years before residents of some of the world's largest cities can be assured of clean air.

In Mexico City, which currently faces extremely high pollution levels, taxis and buses are being replaced or updated with new engines, new cars use unleaded gasoline, and catalytic converters became standard equipment on new cars in 1991. With a

# Designing an Energy-Efficient Car

population of over 20 million people, Mexico City is congested with vehicles. The situation is worsened by the city's location. At 2,240 meters (7,347 feet), the atmosphere contains less oxygen, so fuels do not burn as efficiently and they produce more ozone than at sea level. In addition, the city is surrounded by mountains that trap the polluted air near the ground during temperature inversions that are more common in the winter months. Usually these inversions last only a few hours, but people fear longer inversions, like the one that killed 20 people in the town of Donorra, Pennsylvania in 1948 or the fog that killed 4,000 people in London in 1952.

Some major cities in the United States have experienced significant reductions in air pollution over the last few decades due to pollution control procedures. Yet in 1994, Los Angeles exceeded federal air quality standards for ozone on 88 days and for carbon monoxide on 24 days, while Houston exceeded the standard for ozone on 15 days. Reducing pollution from automobiles will have to be included in any plans to further clean up our cities' air.

## Questions

1. Why does Ricardo wear a surgical mask?

2. What are some of the major causes of air pollution in Mexico City?

3. Assuming that cars are a major source of pollution, make a preliminary proposal for preventing the conditions described in the future in the story. Then describe what you would need to do to find out if your plan is practical.

4. If car designs and energy efficiency remain the same, do you think that large American cities will suffer the same fate as described in the story? Explain your answer.

*Designing an Energy-Efficient Car*

# 58

## Introduction — Automobiles: Then and Now

With the development of the gasoline-powered engine, automobiles became the main source of transportation for most Americans. As gasoline prices soared in the early '70s, manufacturers were faced with the challenge of designing a more energy-efficient car.

## Challenge

Contrast the 1965 model Cadillac with the 1995 model. Think about what may have caused the differences between the two.

Above: 1965 Cadillac Sedan deVille

Left: 1995 Cadillac Seville Touring Sedan

Activity 58

# Designing an Energy-Efficient Car

Cadillacs: Then and Now

|  | 1995 Seville Touring Sedan | 1965 Sedan de Ville |
|---|---|---|
| Cost | $ 45,057 | $ 8,589 |
| Length (in.) | 204 | 224 |
| Curb weight (lbs.) | 3,869 | 4,993 |
| Mileage (mpg) City Highway Overall | 16.0 25.0 17.0 | 12.2 14.9 13.0 |
| Engine Horsepower Torque* (ft. lbs.) | 300 295 | 340 480 |

*Torque is the amount of work needed to turn the drive shaft

## Questions

1. Summarize the differences between the 1995 and 1965 Cadillacs.

2. Why do you think these changes were made?

# Designing an Energy-Efficient Car

## Introduction
### Automobile Fuel Efficiencies

Imagine you are about to buy your first car. What model do you think you will choose? A lot of factors will probably influence your decision. One factor you may consider is the cost of buying the car. You may also consider the fuel efficiency in miles per gallon (mpg). The greater the fuel efficiency, the less you will have to pay for gas!

## Challenge

Prepare graphs of the data provided and use the graphs to understand factors that affect fuel efficiency.

## Procedure

1. Work with your group of four to graph the data on the following page. One pair should make a graph comparing the miles per gallon of each car to its weight. The other pair should compare the miles per gallon to the horsepower. Plot miles per gallon on the y-axis.

2. Share your graph with the other pair in your group as you answer the questions.

*Activity 58*

# Designing an Energy-Efficient Car

*Weight, Horsepower, and Mileage Data for Selected Automobiles*

| Car Name and Model | Weight (lbs.) | Horsepower | Miles/Gallon ** |
|---|---|---|---|
| Acura Integra* | 2,665 | 142 | 30 |
| Buick Park Avenue | 3,640 | 205 | 20 |
| Buick Regal | 3,445 | 160 | 20 |
| Chevrolet Lumina | 3,395 | 160 | 21 |
| Chrysler LHS/New Yorker | 3,605 | 214 | 20 |
| Ford Aspire* | 2,140 | 63 | 36 |
| Ford Taurus | 3,516 | 145 | 21 |
| Geo Metro* | 2,065 | 55 | 29 |
| Geo Prizm* | 2,510 | 105 | 33 |
| Honda Accord | 3,255 | 130 | 21 |
| Honda Civic | 2,440 | 106 | 31 |
| Lincoln Town Car | 4,055 | 210 | 16 |
| Nissan Altima | 3,050 | 150 | 23 |
| Oldsmobile 98 Regency | 3,640 | 205 | 19 |
| Oldsmobile 88 | 3,470 | 205 | 19 |
| Saturn SL | 2,405 | 100 | 29 |
| Suzuki Swift | 1,845 | 70 | 29 |
| Toyota Corolla | 2,540 | 100 | 30 |

*manual transmissions      **Based on EPA ratings and data in Consumer Reports, April 1996

### Questions

1. Describe the trends (patterns) shown by your graphs.
2. Which one of these cars would you choose to buy? Why? What more would you like to know before deciding?

# Designing an Energy-Efficient Car

## Introduction

### Energy Efficiency of Modern Cars

Automobiles have rapidly changed in style and energy use since the earliest models. But, as you can see from the graphs on the following pages, their gasoline efficiency has made little progress since 1989. Why might this be happening? How can we improve cars and make them more efficient?

## Challenge

After discussing the following information with your group and family, develop a list of the most important things to consider if you were to design a new car, the "car of the future."

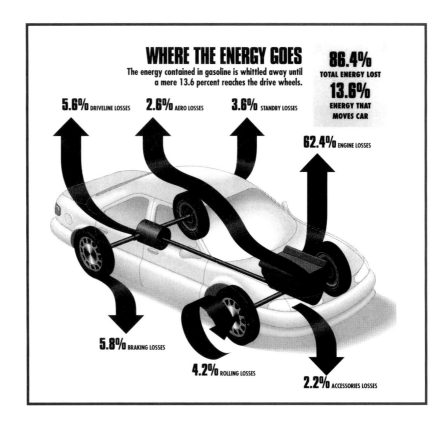

# Designing an Energy-Efficient Car

*Passenger Car Efficiency in Miles per Gallon*

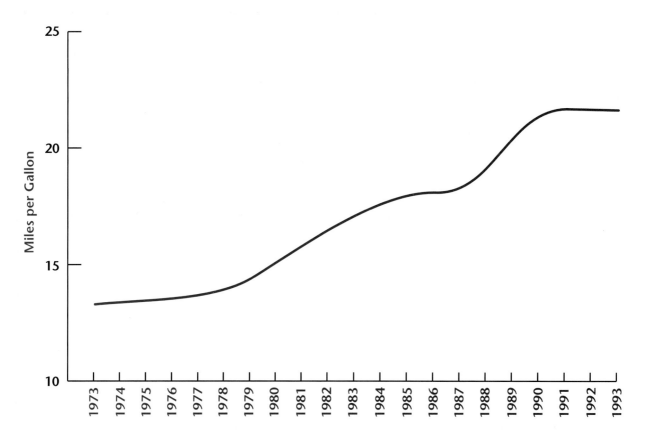

Sources: U.S. Department of Transportation, Federal Highway Administration, Federal Highway Statistics Division

## Designing an Energy-Efficient Car

*Changing the Efficiency (Fuel Economy) of the 1995 Car*

| Change | Approximate Effect on Efficiency |
|---|---|
| Adding an air conditioner | reduces efficiency by 9% |
| Using an automatic rather than manual transmission | reduces efficiency by 6% |
| Using steel-belted radial tires | increases efficiency by 10% |
| Reducing driving speed from 65 to 55 mph | increases efficiency by 5% |
| Reducing the vehicle weight by 500 lb. | increases efficiency by 15% |

# Designing an Energy-Efficient Car

*Energy Sources—Pluses and Minuses*

| | "Clean" Gasoline | Ethanol | Compressed Natural Gas | Electricity | Hydrogen |
|---|---|---|---|---|---|
| **Description** | Today's gasoline combined with oxygen-giving compounds. | Alcohol fuel made by fermenting corn and sugar cane. | Used for cooking and heating in homes—methane. | Derived from recharging a variety of new, high-powered batteries. | Solar energy would be used to split water molecules into hydrogen gas. This would be used in a fuel cell in the car to create electricity. |
| **Advantages** | Reduces air pollution from all cars that use it. Can be distributed now in present gas stations. Doesn't require engine changes. | Has high octane (105) compared to regular gas at 87. Comes from a renewable resource. Has lower pollutant levels (carbon monoxide) than gasoline. | Is abundant in the U.S. Lowers pollutant levels emitted by 60% on the average. Has distribution system already in place. | Has quiet engine. Emits almost no pollutants at the point of use. | Comes from a renewable energy source. Emits practically no pollutants. Produces no carbon dioxide. |

Activity 58

## Designing an Energy-Efficient Car

| | "Clean" Gasoline | Ethanol | Compressed Natural Gas | Electricity | Hydrogen |
|---|---|---|---|---|---|
| **Disadvantages** | Benefits to the environment are not known. Uses a non-renewable resource. Does not reduce carbon dioxide emissions. | Produces less energy per gallon than gasoline. Requires frequent fill-ups. Would use 40% of U.S. grain harvest to produce only 10% of the fuel needed. | Requires heavy fuel tanks. Requires refueling every 100 miles. Takes two to three times longer than gasoline to refuel. Comes from a non-renewable source. | Technology will not be widely available for 10–15 years. Requires over 1,000 lbs. of batteries that need to be recharged for 6–8 hours daily; range is limited to 100 miles. The pollution would be transferred to the source of electrical generation. Speed limited to 30–65 mph. | Technology will not be available for 20 years. Cannot match the performance of today's cars. No distribution system is in place to ship the gas. There are safety issues of hydrogen gas explosions. |
| **Costs** | 12–14 cents more per gallon than present unleaded gasoline | Twice the cost of gasoline | At the moment inexpensive—70 cents per gallon | Cost of electricity to recharge batteries is low, but batteries must be replaced every 30,000 miles at a cost of $2,000 each. | Very expensive and not currently produced on a large scale |

Activity 58

## Designing an Energy-Efficient Car

 **Questions**

1. Collect an advertisement from a newspaper or magazine about any new car. What reasons do you have for buying or not buying this car? Be prepared to share your reasons in class.

2. Describe the most important changes you would make in order to design a new car that is more energy efficient and environmentally friendly. Include materials you would use and why, energy used by the engine and the fuel it might use to power it, and the effect of the car on the environment (including effects on human health). Be sure to describe the evidence and tradeoffs you used in making your decisions.

# Designing an Energy-Efficient Car

**58**

## Introduction

### The Hyper Car

In 1993, leaders from government and the automobile industry held an historic conference. Together they proposed the goal of manufacturing a super efficient car by the year 2003.

The proposed car will be environmentally friendly and could obtain a fuel economy of nearly 80 miles per gallon of gasoline.

## Challenge

Read about the Hyper Car and compare it to the car of the future that your group has been designing.

The Ecostar has been produced as part of Ford's effort to research and develop an all-electric vehicle. Similar efforts are under way by other manufacturers.

Activity 58

# Designing an Energy-Efficient Car

Materials scientists, automobile engineers, energy specialists, and other experts are developing a car completely different from the cars of today. This car is not just a simple modification of present designs; instead it is a complete breakthrough in material and energy use. It is called the Hyper Car.

Cars lose energy through heat loss. Think of a dollar of energy going in the tank and 14 cents coming out as usable energy. The rest is wasted! But if new thinking about materials usage, energy storage devices, and aerodynamic design, coupled with high-tech manufacturing processes, are applied to this problem, a truly remarkable vehicle can be produced.

Most of the present car's weight comes from the steel body and the high-horsepower engine. The engine's power is used to accelerate this heavy weight from a stopped position (0 mph) to a cruising speed of 55 mph. Once the car reaches 55 mph, only 8 horsepower of energy from the engine are needed to keep it moving. With the introduction of synthetic materials—carbon fibers that are stronger than steel—the car's body weight can be cut from nearly 3,200 pounds on the average to 1,400 pounds. This change alone has a big impact on mileage. It means that the engine can be much smaller and can require fewer horsepower which would result in greatly increased fuel mileage. In fact, the engine would need to be only slightly larger than the one on a large lawnmower!

In the Hyper Car, gasoline or an alternative fuel would still act as the primary fuel source. But the engine would create electricity that could be used by direct-drive electric motors attached to each wheel. These motors would power the car. When the car slowed down or braked, the engine could also function as a generator of electricity to reclaim the car's braking energy losses.

# Designing an Energy-Efficient Car

By molding the frame and body into a single unit made of carbon fibers, the Hyper Car design would minimize losses in energy due to air resistance. However, these fibers currently cost $18 per pound compared to 50 cents per pound for steel. The car body would cost about five times more than today's cost of a steel body, but it would allow cars to be manufactured and customized to a customer's specifications within a matter of days. Think about designing your own car body and seeing it delivered a week later!

The car that is envisioned would require a new kind of engine, battery, and materials technologies that are still in development. A car like this would be beyond the reach of most Americans' pocketbooks now. It would also be smaller, and all of us would need to change the way we view cars. But a car that can cross the country on one tank of gas may not be that far in the future.

Manufacturing the Hyper Car at a reasonable cost will be a great challenge. However, meeting this challenge is likely to help solve other problems. The new technological advances and materials developed in producing the Hyper Car are likely to have many other uses, not only for cars but in our homes and lives.

## Questions

1. Compare your car design with the Hyper Car. What new ideas would you like to add to your design?

2. What if the Hyper Car were made so cheaply that every person in the world could own one? Write 3–4 paragraphs explaining your reaction to a world with as many cars as people. Include your ideas about alternatives to this future.

Activity 58

# Extension: Designing an Energy-Efficient Home

## Introduction

In Activities 49 and 51, you investigated electrical energy use in your home and made a plan for reducing your electrical energy usage. But there are other ways to approach energy efficiency in the home. Today many architects and home builders are turning to energy-efficient designs and solar power to help reduce the use of electricity and gas for home heating and cooling. Homes are kept comfortable through the use of a variety of approaches, including architectural design elements, selection of building materials and techniques, location and orientation of the home relative to the sun and the site, and passive solar heating and cooling.

## Challenge

Design an energy-efficient house for your local climate that meets the needs of your family. The house should be warmed by the sun in winter but not overheated in summer. It may also take advantage of other energy-efficient methods you have learned about in this part of the course.

# Extension

## Designing an Energy-Efficient Home

### Background | Passive Solar Energy and the Home

Energy-conscious design is a growing field in the building profession. Taking advantage of the sun, climate, and local materials to provide a comfortable dwelling is not a new idea. Consider the adobe used to construct the thick-walled Spanish missions in the Americas, the cliff dwellings of Native Americans in the Southwest, or the underground homes built in the desert of Australia. Each of these types of home takes advantage of materials and design principles to provide a comfortable living environment without relying on mechanical heating and cooling equipment.

Passive solar heating can be accomplished simply by leaving the south side of a house unshaded during the winter, by using wall materials that store heat, or by building sunrooms. Passive solar cooling can be accomplished by ensuring that there is good ventilation and by providing shade during the hot summer months. Additional strategies for energy-efficient homes may take advantage of the waste heat from incandescent lighting systems and other appliances or minimize the heating effects associated with their use, depending upon the climate and season. Solar heating can also be used to provide hot water as was introduced in Activity 56. In Activity 57 you used some of the new kinds of window coverups that control transfer of solar energy. Think about how you can use these ideas in your planning.

For further information, you may wish to contact your state energy office or the U. S. Department of Energy's Energy Efficiency and Renewable Energy Clearinghouse at 1-800-363-3732.

## Designing an Energy-Efficient Home — Extension

### Procedure

1. Prepare a drawing of an energy-efficient house for your local climate that meets the needs of your family. Label your drawing. You may wish to write a brief description of your design and why you think it will reduce the use of electricity and/or gas (approximately 200–300 words). Be sure your design uses some of the approaches you have learned about in this part of the course. You should also include each of the following:

   - **Climate**
     Include the sun's position and seasonal movements, wind direction, and variations in temperature and humidity.

   - **Orientation of the house**
     Keep in mind the direction that each of the rooms faces.

   - **Windows and shading**
     Describe the direction of major windows and the use of shades, eaves and/or landscaping.

   - **Building materials**
     Choose materials that absorb warmth from the sun and can release the warmth gradually during the winter, and be sure to provide insulation against both summer heat and winter cold.

2. If you have time, members of the class can make models of their designs and test them in the sun or with a light bulb as a simulated sun.

# Part Four

# Environmental Impact

# Part 4

# Table of Contents

**Introduction** .................................................................................................. 1

59. Industry Comes to Town ........................................................................... 2

60. Pinniped Island ........................................................................................ 18

61. Synthesis, Testing, and Redesign of White Glue ..................................... 22

62. Scaling Up Glue Production .................................................................... 27

63. Packaging and Labeling Products ........................................................... 30

64. Planning a Factory ................................................................................... 32

65. An Island Parable ..................................................................................... 46

# Part 4: Environmental Impact

## Introduction

In this final part of *Issues, Evidence and You*, you will have an opportunity to apply your learning from earlier parts of the course to a new issue—whether or not to build a factory in an island community. People are probably making decisions like this in your community right now. They may be deciding whether to build industrial plants, shopping malls, a housing development, or a new transit system.

When a community is considering a major development project, the agencies involved usually prepare an *Environmental Impact Report*. The decision about whether an Environmental Impact Report is required is made at the local level when there is evidence that the project will have a significant impact on the environment. The purpose of the report is to tell the public how the project will affect the environment and what steps will be taken to *mitigate* (reduce or compensate for) the impact. The benefits of the project are also reported. The decision about whether to approve a project is not based on a threshold of impact, but rather on an overall balance of the impacts and benefits of the project. In other words, if the benefits outweigh the negative impacts, the project may still be approved.

You will assume the role of a member of a team that is trying to plan a factory that will benefit the island community with as little negative impact on the environment as possible. You will first learn about the factory's product, including its method of production and the wastes it generates. Each team will prepare a factory proposal, including an Environmental Impact Report, for consideration by the island's residents—your classmates. Your report will include your plans for obtaining the water, materials, and energy needed by the factory, disposing of the factory wastes, and maintaining water quality and the overall environment of the island. When you have completed your reports, the class will decide whether to approve one of the factory proposals.

# Activity 59

## Industry Comes to Town

### Introduction — Comparing Four Industries

You will investigate how members of a community can compare the possible impacts of four different industries. You will begin by rating four industries based on what you already know about them. Then you will gather more evidence about the products and by-products of each industry before reconsidering your ratings.

### Challenge

Rate four industries according to the categories in the chart. Identify the advantages and disadvantages of locating each industry in your community.

Tomato processing plant

# Industry Comes to Town

## Industry Descriptions

### Chemical manufacturing

Many chemical companies specialize in certain kinds of chemicals. Consider a chemical industry that makes plastics for a wide range of products. The raw materials for making plastics come from processed crude oil. The plastic produced is usually shipped to other factories to be manufactured into final products.

### Computer production

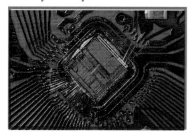

Manufacturing computer circuit boards requires special plastics, cleaning solvents, and other materials, such as acids. Various chemical processes are used to produce the circuit boards.

### Food processing

Raw tomatoes are brought to processing plants where they are washed, crushed, cooked, and made into a variety of tomato products. They are packaged for shipment to markets. The process requires the use of some chemicals in addition to the natural ones in the tomatoes.

### Oil refining

Crude oil comes in by pipeline and is processed to produce gasoline, other fuels, and raw materials used for producing plastics.

# Industry Comes to Town

## Procedure

### Part One: Comparing Four Industries

1. Four industries are listed and are briefly described on the previous page. Imagine that they want to locate in or near your community.

2. Make a chart in your Journal similar to the chart below.

3. Rate each kind of industry on your chart, using the 1–5 scale for each pair of word opposites. For example, in section A, if you think an industry is good, mark a 5 in the appropriate space. If you think an industry is bad, mark a 1 in the space. Use a 3 if you think the industry is neither good nor bad. Use a 2 if you think it is somewhat bad or a 4 if you think it is somewhat good.

4. Base your ratings on the information about each industry given on the previous page and on anything else you know about the industry.

| | Category | | | | | Ratings | | | |
|---|---|---|---|---|---|---|---|---|---|
| | 1 | 2 | 3 | 4 | 5 | Chemical Manufacturing | Computer Production | Food Processing | Oil Refining |
| A | bad | | | | good | | | | |
| B | ugly | | | | beautiful | | | | |
| C | worthless | | | | valuable | | | | |
| D | dirty | | | | clean | | | | |
| E | harmful | | | | helpful | | | | |
| F | hazardous | | | | safe | | | | |
| G | unnecessary | | | | necessary | | | | |
| | | | | | My totals | | | | |
| | | | | | Our group's average | | | | |
| | | | | | Class average | | | | |

Activity 59

# Industry Comes to Town

## Procedure

### Part Two: Issues Related to Siting Industry in a Community

1. Imagine that your community is trying to decide whether to encourage new industry to come to town. A citizens' committee has asked you to join them to review the advantages and disadvantages of bringing new industry to the community. In preparation for the committee meeting, make a chart of all the advantages and disadvantages you can think of in locating an industry in your community.

2. Compare and discuss the charts of all members of your group. Reach a consensus on identification of the three most important issues to resolve before deciding whether an industry can come to your community.

3. As a class, come to agreement on the three issues considered most important by your class. Record them in your Journal.

# 59 Industry Comes to Town

## Introduction | Evaluating Industry Reports

Your group will evaluate detailed reports submitted by four industries that want to locate in your community. Each group member will produce a summary about one industry and share the results with the group.

## Challenge

Read the report submitted by each industry. Separate the evidence from the company's opinions about their operation and safety.

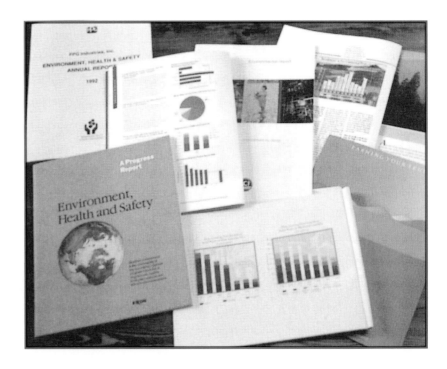

# Industry Comes to Town

## Procedure

1. Read about your assigned industry. As you read, try to separate evidence from opinions.

2. In your Journal, prepare a summary about the industry that includes answers to the following questions.

    a. What are the raw materials (reactants) used in the industry?

    b. What are the various products of the industry?

    c. What are the by-products of the industry? Are they reused or disposed of as wastes?

    d. What are three advantages of having this industry in the community?

    e. What are three disadvantages of having this industry in the community?

3. Share your results with the other members of your group.

# Industry Comes to Town

## Industry Overview: Kelly Chemicals

> **Working for You**
>
> Kelly Chemicals owns and operates 15 plants around the country. These plants produce many chemical products that most people use daily in one form or another but may not know by name. Kelly Chemicals makes substances that are used in textiles, lubricants, tires, plastic packaging, detergents, adhesives, and cosmetics. Kelly Chemicals would like to build a plant in your community, providing employment and needed products for the local economy.

### Making Chemical Building Blocks

The proposed plant's main products—olefins—are compounds of carbon and hydrogen made from natural gas. They are sometimes called "building blocks" because they must be processed further before they can be turned into finished products. Some of the major types that will be produced at this plant are ethylene, propylene, polyethylene, and polypropylene. These molecules are the raw materials used to produce many common consumer products. Other chemicals made at the plant will be sold to other chemical plants for additional processing into such items as dry-cell batteries, plastics, detergents, adhesives, and synthetic oils.

### Community Concerns

We know that there are many concerns regarding the operational safety of the plant. Other plants of similar design and operation have earned numerous safety awards. Such plants have chalked up more than 2 million continuous people-hours without an accident serious enough to require time off from work.

#### Fire and Explosions
The olefins, and the hydrocarbons used to make them, are flammable. Sophisticated electronic and mechanical means will be used to monitor operations for signs of problems. The ethylene

# Industry Comes to Town

unit alone will have 1,300 sensors that will relay information continuously to a control room where engineers can, if necessary, shut down operations in the plant. We will regularly test our detailed emergency procedures, which will be developed in cooperation with local community leaders. Employees will be required to keep their skills sharp with regular training sessions and updates.

*Waste Issues*
Kelly Chemicals has a policy of waste minimization. Plant managers work with engineers to develop processes that minimize the waste generated by plant operations. We register all solid waste by-products with appropriate authorities and dispose of them in compliance with all federal, state, and local regulations.

We treat waste water from plants, like the one proposed for your community, to a level of quality far higher than required by state laws. In fact, in similar plants, up to 2 million gallons of water per day come out cleaner than before treatment. At one of our other plants, processed waste water from the plant supplies a well-stocked artificial lake rich in native waterfowl, fish, and other animals. A similar project will mitigate (compensate) for the impact of the proposed factory. We will construct an artificial lake near the proposed factory site that will provide sanctuary for local birds affected by the industrial area.

## A Community Benefit

The many uses of olefins ensure their continued importance to the economy. The plant will provide jobs for more than 700 people. The plant will contribute millions of dollars in local, state, and federal taxes and many more times that in local salaries.

Kelly Chemicals believes it can operate a plant that is safe for workers and community residents alike. State-of-the-art technology and careful waste minimization procedures will combine to allow the plant to operate without excessive impact on the environment.

## Industry Overview: Midvale Computers

> **Working for You**
>
> Midvale Computers is among the fastest-growing companies in the field, rapidly opening new plants across the country. The plant in your community will produce circuit boards for personal computers. Circuit boards are the backbone of the computer, on which other parts, such as memory chips and microprocessors, are installed.

### Making Circuit Boards

Circuit boards are made from rigid plastic sheets covered with a thin metal coating. Before the coating is applied, the plastic surface is cleaned with various liquids. These liquids dissolve dirt and other impurities, such as oil and grease, that do not dissolve in water.

The lines of metal that conduct electricity on the circuit boards are produced by a process called *etching*. Acid is used to etch (dissolve) large portions of the metal coating, leaving only ribbonlike lines of metal on the plastic. Once finished, the boards are cleaned again with solvent sprays; some parts must also be painted or coated with various materials.

### Community Concerns

Community concerns about the operations of our plants are usually directed toward problems associated with the acid, metal, paint, and solvent by-products generated by the production of circuit boards.

# Industry Comes to Town

*Acid and Metal Wastes*

The etching process produces an acidic waste with a high concentration of dissolved metals. These wastes cannot be disposed of without treatment. We neutralize the acid by adding measured amounts of base. We remove the metals by adding chemicals that combine with the metals to form solids. These solids are easily removed from the waste stream by filtering. We then discharge the treated wastes to the city sewer system.

The metal solids that are filtered out are further treated to recover valuable metals for reuse. We pump the remaining nonvaluable metals into a large, lined holding pond to settle. As they settle, they form a very thick material called *heavy metal sludge*. This sludge is classified by the Environmental Protection Agency (EPA) as a hazardous waste because it can be dangerous to inhale or ingest, even in small amounts. According to federal law, this heavy metal sludge must be removed to a special landfill classified for hazardous wastes. The proposed plant will produce approximately two truckloads of the sludge per month.

*Paint and Solvent Wastes*

The cleaning solvents can be recycled and reused a number of times before disposal. Paint and solvent wastes that cannot be reused are also classified as hazardous by the EPA and must be disposed of safely. This plant will produce one truckload of paint and solvent wastes for disposal per month.

*Groundwater Contamination*

Recently, there has been extensive groundwater contamination associated with some computer companies. In most cases, this contamination is a result of leaking underground storage tanks. All chemicals used at the Midvale plant will be stored in

*Continued on next page*→

aboveground, double-walled tanks on paved floors and checked constantly for leaks. While accidental releases occasionally happen, these safety features help prevent chemicals from leaking into the ground, where they might contaminate the groundwater.

*Worker Safety*

Midvale Computers, recognized in many communities for its achievement in worker safety, has one of the lowest accident rates in the field due to strict safety procedures. These include safety training for new employees; use of safety eyewear, gloves, and other protective clothing; reduction in the number of employees who must come into contact with the chemicals; and a worker health-maintenance policy, which requires yearly physicals.

## A Community Benefit

Midvale Computers is a successful company that produces quality products safely and with very little impact on the environment. No smokestacks, loud machinery, or other signs of an industrial presence will be noticeable from the roadway. In fact, the beautifully-landscaped plant site will almost pass for an office complex or medical facility.

As is the case with any industrial development, the plant will place additional demands on the town's water supplies and sewage treatment system. Current predictions made by the company indicate that the town will have no trouble meeting its other needs while still providing for the needs of the plant.

Your community will benefit from the local taxes Midvale Computers will pay. Midvale means jobs and a healthy economy for your community.

# Industry Comes to Town

## Industry Overview: Red Ripe Foods

> **Working for You**
>
> The Red Ripe Foods Corporation operates a large number of factories across the country. Although the company produces a wide variety of fruit and vegetable products, it is best known for its tomato products—tomato paste, juice, sauce, and catsup. Our plans are to build a tomato-processing plant in your community.

### From Field to Can to Grocery Store Shelf

After tomatoes are harvested in the fields, they are shipped in large trucks to the factory. We unload the tomatoes into storage tanks filled with water. Rollers move the tomatoes from the water bath through a series of sprayers. The combined action of the rolling and spraying gently loosens surface dirt.

The cleaned tomatoes are then transported to an area where they are trimmed and visually inspected. Next they pass briefly through a steam bath, which causes the skin to loosen. Metal hoods above the steam bath release the steam harmlessly to the outside atmosphere so that it does not disperse throughout the plant, where it might be hazardous to plant workers. Pumps move the tomatoes to the pulper, which separates the skins and seeds from the tomato pulp. The seeds are collected, dried and packaged to be sold; the skins must be discarded as waste. We then pump the tomato pulp through glass pipes to be processed into juice, paste, sauce, or catsup.

### Community Concerns

All industrial food processing requires attention to safety and the local environment. In order to limit the impact of plant smells and noise, we will site the proposed plant on the outskirts of your community. The trucks will load and unload at off-peak hours to minimize traffic congestion and the risk of accidents. As is the case with many agricultural operations, work at the plant will be highly seasonal, depending on the availability of ripe tomatoes.

*Biological Hazards*

We strictly monitor the chemical environment throughout every phase of production to control bacteria that can spoil food. As part of the canning or bottling process, we pasteurize the paste, juice, sauce, and catsup to kill microorganisms. The plant machinery is constantly supervised and cleaned frequently to minimize the possibility of bacterial growth. All waste water passes through a fine mesh screen to remove tomato solids. The solids are sold as animal feed or as fertilizer for other crops. The small amount of tomato solids that cannot be reused must be disposed of as solid waste in the area landfill.

Wash water may contain some bacteria or other microorganisms washed from the tomatoes in the early cleaning stages. We treat this water, and all other water used at the plant, before release.

*Pesticide Contamination*

We test all tomato solids, as well as all water used in the unloading and cleaning steps, for pesticide residues before disposal. Growers must register the pesticides used with local and state authorities and must provide this information—as well as dates of application—with each shipment to the plant. Red Ripe Foods does not accept produce from growers unless a certain period of time has passed, as required by law, since pesticides were last used on that crop.

### Community Benefit

Red Ripe Foods would like to open a plant in your community. The plant will provide employment opportunities for local residents, as well as food products that can be sold locally or shipped elsewhere. The management of Red Ripe Foods would like you to know about what goes on at the factory so that you can make an informed decision regarding the plant. Red Ripe Foods believes it can operate a plant that will provide jobs and needed products for the local economy with minimal environmental impact.

# Industry Comes to Town

## Industry Overview: Star Refineries

> **Working for You**
>
> Star Refineries operates a number of facilities across the country. Its plants process crude oil to make many different kinds of products. Some of these products you may know and use. They include fuels for automobiles and aircraft; oils for lubricating cars; heating oils for homes; asphalts for roads; and the substances required by chemical companies to make many other products used in fields as diverse as agriculture, sports, and medicine. Star Refineries wants the public to know about its operations so that residents of the community will be able to make informed choices about the plant location.

### The Process of Refining Oil

The crude oil processed in the plant was formed millions of years ago in underground pools or in certain kinds of rock formations. It is a mixture of different hydrocarbons, which are molecules containing hydrogen and carbon.

After the oil is removed from the ground, it is transported to the refinery by pipeline, ship, truck, or some combination of the three. Pipelines will transport the oil to the proposed refinery. Once at a refinery, the first stop for the oil is usually large storage tanks where it is stored before processing.

The first step in the processing begins at a distillation tower, which may be many stories high. Here, the crude oil is separated into the different kinds of hydrocarbons. This separation process works by boiling the crude oil. Lighter hydrocarbons, with lower boiling temperatures, rise to the top of the tower and heavier hydrocarbons, with higher boiling points, are found at the bottom of the tower. Gasoline boils first so it rises to the top in gaseous form. It is then drawn off and cooled, which causes it to become liquid again. Heavier hydrocarbons, such as those that make up heating fuels, collect in the middle of the tower, where they are removed. The heaviest components concentrate at the bottom of the tower, where we collect them for further processing into asphalts and the basic ingredients for the manufacture of a variety of chemical products.

After the crude oil is separated, each part must receive further treatment before use. About one-third of a barrel of crude oil processed by Star Refineries ends up as gasoline. The rest becomes jet fuels, kerosene, lubricants, thinners, solvents, and many other products.

## Community Concerns

Communities in which refineries are located are generally concerned about the visual impact of the facility, the risks of fire and explosion, the effects of oil leaks or spills on the environment, and the effects of the refining process on air quality. Star is an industry leader in the development of procedures to prevent accidents and minimize environmental impact.

### Visual Environment
The plant will be located in the existing industrial area near the river north of town, far from any residential areas. Star Refineries will construct a parklike border of trees around the plant and along the opposite shore of the river. This will help absorb noise and reduce (mitigate) the visual impact of the refinery.

### Fires and Explosions
Our procedures to prevent fire and explosion include constant checking of storage tanks and plant equipment for any signs of leaks or other problems. To minimize the risk of fire, the large storage tanks used for the storage of oil prior to processing are located a great distance away from other plant equipment.

# Industry Comes to Town

*Air Pollution*

Techniques for preventing the release of airborne pollutants have improved greatly. We collect the hydrocarbon vapors and burn them off cleanly, producing carbon dioxide gas, water, and a little smoke. The carbon dioxide gas has the same chemical composition as the carbon dioxide in your breath and is nontoxic in the concentrations produced by the plant. The distillation towers contain filtering devices that remove particles from the refinery exhaust. These hydrocarbon-containing wastes are either recycled or treated. One treatment method is to use oil-consuming bacteria that break down hydrocarbons to release carbon dioxide gas and water and create organic material that can be packaged and sold as a soil additive.

*Water Pollution*

The refining process requires millions of gallons of water daily, much of it to keep the towers cool. We recycle most of the water and reuse or treat it to render it nontoxic before releasing it to the municipal waste-water system.

## A Community Benefit

Star Refineries would like to build a plant in your community. The management of Star works closely with government regulators to maintain its excellent record and its reputation as an industry leader in safety and environmental issues. Star developed many of the standard industry procedures for the reduction of dangerous emissions and recovery of wastes. Star Refineries will bring many good, safe jobs to your community.

# Activity 60 Pinniped Island

## Introduction

### Pinniped Island and Its Future

Imagine you live on Pinniped Island and are concerned about its future. The lack of good year-round jobs causes many young people to leave the island. Your interest in the island's future makes you want to know more about the island and its resources.

## Challenge

Read about Pinniped Island and its current resources. Think about the advantages and disadvantages of bringing new industry to the island.

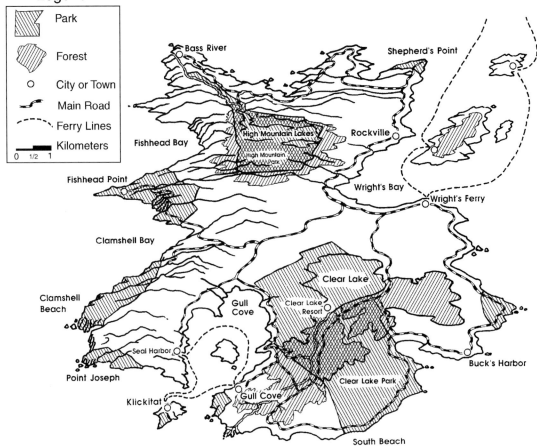

## Pinniped Island

Pinniped Island is located somewhere in the ocean. Irregular in shape, it measures roughly 20 miles long and 15 miles wide and is fairly mountainous. The highest mountains are in the northwestern part of the island. Mount Reyes, in the High Mountain Park area, is the highest peak—5,253 feet. The eastern side of the island is hilly, with no peaks above 3,000 feet.

Much of the island's yearly rainfall of 25–30 inches is captured in the island's many small streams and lakes. Clear Lake, which is located in the south central region, serves as the island's main reservoir. High Mountain Lake is a smaller lake in the northwestern region of the island that is isolated from the main river system.

The island's mild year-round climate produces average daytime temperatures between 5° and 10°C (approximately 40° and 50°F) in winter and 20° and 30°C (approximately 68° and 87°F) in summer. Winds are generally from the northwest and can be quite brisk on the western side of the island.

Sheep and cows graze the island's extensive grasslands. Forested areas are limited, and large trees are sparse except near the lakes and rivers. The island does not have any minerals of economic importance.

Many of the island's 15,000 people live in small towns along the coast. Wright's Ferry (population 4,200) and Buck's Harbor (population 2,900) are the largest towns. Tourism, fishing, raising sheep for wool, and dairy farming are important to the local economy. Travel to the mainland is by ferry; the express takes approximately 1 hour, while the ferry that stops at neighboring islands takes 1 hour and 45 minutes.

# Pinniped Island

## Introduction

### Island Clarion

Last night the Island Council discussed locating a glue factory on the island. Some members of the council think a glue factory would improve the economic outlook on the island. Today the following comments were printed in the paper in answer to the question: Are you in favor of the factory?

## Challenge

Read each response carefully. In each response, what is based on evidence and what is based on personal opinion? What other evidence would you like to have to help you make a decision about the factory?

*Pinniped Island*

# Island Clarion

**What's your opinion?** *Are you in favor of building a glue factory on Pinniped Island?*

**Jacob Smith,** self-employed concrete contractor
"Sure, I'm all for it. There's not enough work on this island, and a glue factory will mean a lot of work building roads and factory buildings. Unemployment on the island is nearly 20% and even higher in the winter. I heard that the factory will provide nearly 200 jobs. This is an important step in improving our economy."

**Cameron Davis,** shipping clerk
"Think of all the jobs it will create, not only in the factory, but in services needed to support it. Our shipping and cargo company will certainly increase the number of shipments that are transported both to and from the island."

**Melinda Phipps,** school guidance counselor
"Yes. I'm tired of seeing students from our high school leave to get jobs on the mainland. It's better for them to work here and remain part of the island community."

**Alana Teagues,** computer programmer
"No. The factory will destroy the environment and only help the fat cats who operate it. All factories are dirty and noisy, and you know how they create pollution."

**Isaac Roberts,** middle school student
"Maybe. I don't know. It depends on a bunch of things. It's difficult to decide without knowing more."

**John DiSarro,** high school senior
"Yes, because then I can get a job and pay for my car insurance."

**Hans Efger,** fisherman
"No. The fish populations are decreasing, and the waste from a factory would destroy the fishing industry on this island."

**Raylene O'Malley,** Island Utility Department
"No. Our energy analysis shows that we have enough power to meet current needs, but a factory will use a large part of our reserves. We will need to build another power plant, with money we don't have, if the island continues to grow and needs more power."

**June Chan,** neighbor of existing dairy
"No. The factory will need more milk if it's to grow and make money. The dairy I live next to is a source of flies and foul smells. There will be more trucks coming and going to ruin our rural atmosphere."

**Fernando Castillo,** proprietor, The Island Cafe
"I don't know yet. It's true that our unemployment rate is a problem. I would be concerned that a factory would cause pollution and traffic problems. I'm waiting to find out more before I make up my mind."

# Activity 61 Synthesis, Testing, and Redesign of White Glue

## Introduction

### Producing White Glue

In order to evaluate the impact of a glue factory on the island, you will need to gather evidence about the products and by-products of glue production. When milk is warmed or acidified, most of its protein content precipitates out as *curds*. What is left is a solution containing milk sugars and water soluble proteins called *whey*. The curds from curdled milk, called *casein*, produce a form of glue that is very strong and water resistant.

## Challenge

Working with your partner, use the materials and procedures supplied to make, and then test, white glue. Redesign your approach to produce a higher quality glue while minimizing the amount of waste.

# Synthesis, Testing, and Redesign of White Glue

## Materials

*For each group of four students:*

- One 30-mL dropping bottle of 5% household vinegar
- One 30-mL dropping bottle of 5% household ammonia
- pH paper or Universal Indicator solution
- Materials to glue together—small pieces of paper, wood, plastic, aluminum foil, glass slides—as available

*For each pair of students:*

- One SEPUP tray
- One SEPUP filter funnel assembly
- One graduated container
- 10 mL of warm (40°–50°C) milk
- One piece of filter paper
- One stirring stick
- Water
- One paper towel
- One dropper
- Containers for storing glue

# 61
## Synthesis, Testing, and Redesign of White Glue

## Procedure

### Part One: Preparation of White Glue

 **Safety**

Be sure to wear safety eyewear during this investigation. Take care when handling hot liquids.

1. Before you begin making the glue, plan the two tests you will use to determine the quality of your glue.
2. Add 10 mL of warm milk to the graduated container.
3. Add 12 drops of vinegar to the 10 mL of warm milk. Stir after the addition of each drop.
4. Filter the mixture, separating the curds (the solids) from the whey (the liquid). Put the whey in the waste container.
5. Transfer the curds from the filter paper to the graduated container and rinse them with clean water.
6. Mix the curds with 12 drops of ammonia. Warm the mixture, as necessary, by using the water bath and stirring until all the curds are dissolved.
7. Carefully add more water, a drop at a time, if you think the glue should be thinner.
8. Perform two tests to determine the quality of your glue. Label the glued objects and place them in a safe location until the next class session.
9. Clean the materials thoroughly to remove any traces of glue.

 **Data Processing**

In your Journal, record your experimental procedures for making and testing the glue and your observations. You will share your results with the class tomorrow.

# Synthesis, Testing, and Redesign of White Glue

## Procedure

*Part Two: Testing and Redesign of the Glue*

1. Observe the tests you set up in the previous session.

2. Make a table to record your test results. For each test, you should record the glue characteristic you are testing, the procedure you used, your test results, and conclusions.

3. Discuss your results with your group or your class.

4. Based on the discussion, work with your partner to plan the changes you will make to improve your glue recipe.

## Data Processing

Record your plan for making an improved glue. Be sure to describe why you think your plan will result in better glue.

# 61 Synthesis, Testing, and Redesign of White Glue

## Procedure

*Part Three: Making a New and Improved Glue*

1. Follow your plan to make your new batch of glue.

2. Use the same tests you used before to test the new glue.

## Data Processing

1. Record the procedure you used to make an improved glue.

2. Compare your new glue to the first batch you made. How did you change your glue-making procedure? Did the new procedure make an improved glue?

3. How would you modify your process and/or ingredients to help you maximize the production of white glue while minimizing the amount of wastes produced?

4. Discuss with the other team in your group the similarities and differences in the redesigned approach that each team adopted for making a better glue. Come to a consensus on a process your group will use for the scaling up of production in the next activity.

# Activity 62 Scaling Up Glue Production

## Introduction

### Scaling Up

You have produced a small amount of glue. Now you want to know if there is a market for the product. You will begin by scaling up the production of white glue from experimental lab amounts to a quantity large enough to meet the needs of the agreed-upon test market. Based on your findings, you will plan to scale up your process to make enough glue to supply everyone on the island for a year.

## Challenge

Design a process for scaling up your method for making glue.

Imagine scaling up cookie production from a few dozen homemade cookies to thousands of cookies for a national market.

# Scaling Up Glue Production

## Procedure

1. Working with your group, use the tables on the opposite page to decide on a proposed test market for the glue.

2. Decide how many 60-mL (2-ounce) bottles of glue you will need for test marketing.

3. After discussing the various proposals with other groups in the class, record in your Journal the size and makeup of the agreed-upon test group and the number of bottles of glue needed for the test marketing.

4. Working with the other members of your group, design a plan for scaling up to produce the quantity of glue for the needed test market. Make sure that your plan includes at least:

    - the amount of each ingredient you will use;
    - the approaches to production you will use; and
    - how you will minimize and then handle the wastes produced.

5. Be prepared to present your plan to the class.

6. Based on the reports and discussions about scaling up for test marketing, and the island information given on the opposite page, estimate the amount of glue needed to supply Pinniped Island for one year.

7. Use your estimated market for the whole island to design a plan for making enough glue for the island for one year.

# Scaling Up Glue Production

 **Data Processing**

Produce a report that describes how you made your estimate of the size of the glue market on the island *and* how you used that estimate to scale up the glue recipe. Be sure to show all the steps of your math work. (Other people might not agree on your "market estimate," but don't worry about that. Do the best you can, and *explain your thinking* so others can react to it. Then use the number you came up with to scale up the glue recipe.)

## Island Information

*Population*

| | |
|---|---|
| Children (newborn–20) | 5,000 |
| Adults (21–55) | 8,000 |
| Seniors (over 55) | 2,000 |
| Total | 15,000 |

*Residences*

| | |
|---|---|
| Residences with children | 2,700 |
| Residences without children | 2,300 |
| Total | 5,000 |

*Schools*

| | |
|---|---|
| Day care centers | 4 |
| Preschools | 2 |
| Elementary schools | 5 |
| Middle schools | 2 |
| High schools | 1 |
| Colleges and trade schools | 2 |
| Total | 16 |

# Activity 63: Packaging and Labeling Products

## Introduction

### Packaging the Glue

Based on the homework assignment and the class discussion, you will compare your group's glue package to the commercial one.

## Challenge

Rate and compare the two packages based on a ten-point scale for each characteristic considered important by the class.

## Procedure

1. Produce a table like the one on the opposite page for your group. In the extra spaces at the bottom of the table, include all *other* packaging characteristics the class considers important.
2. Rate your group's package and the commercial one based on a ten-point scale for each category.
3. Record the total score for each package in your Journal.
4. Before discussing your answers with the other members of your group, answer (in your Journal) the three questions listed below the table.
5. Discuss your answers with your group and come up with a consensus.

# Packaging and Labeling Products

*Comparing Your Package to a Commercial Package*

| Categories | Your Package | Commercial Package |
|---|---|---|
| Cost | | |
| Strength | | |
| Can it be reused or recycled? | | |
| Environmental impact as waste | | |
| Convenience | | |
| Labeling | | |
| | | |
| | | |
| | | |
| **Total Points** | | |

### Questions

1. Based on your results above, what kinds of changes could you make to your glue container to help improve its overall rating?

2. What kind of tradeoffs were made by the manufacturer in producing the commercial glue container?

3. Why do you think it is possible or impossible to produce the perfect (ideal) glue container?

# Activity 64
## Planning a Factory

### Introduction

**Plan Your Factory**

The Pinniped Island Council has agreed to consider proposals for construction of a glue factory on the island. Your company must submit its factory plan and a complete environmental impact report. All submissions will be evaluated competitively by the council to decide whether one of the companies will be approved to build a glue factory.

### Challenge

Think of your group as a company and design a factory and total plan for producing glue on Pinniped Island. Prepare an environmental impact report that justifies your company's decisions.

# Planning a Factory

## Procedure

*Part One: The Preliminary Plan*

1. Use the summary information on the following page to help you begin your factory plan.

2. Develop a company (group) statement of your preliminary ideas of how to meet each of the agreed-upon needs for a factory on the island. Reflect on what you have learned throughout the year. Use your Journal and Student Books as resources to help you plan.

3. You may wish to identify several alternatives that you will choose among later. Your company's statement should help you identify additional information you will need to make final decisions and to prepare the environmental impact report.

4. Include rough maps and/or sketches to illustrate your project.

# Planning a Factory

1. **New construction**

   Glue production will require factory buildings. Your plan may also require additional roads, docks, or other construction for transporting the raw materials and finished product. Describe everything that will be built as part of your project.

2. **Material resources: Milk**

   You will need a source of milk. There are three existing dairy farms on the island. A typical farm there covers nearly 28 hectares (70 acres) of land. Each farm has about 500–600 cows on the dairy, or 30–38 cows per hectare (12–15 cows per acre) of pasture land. The number of cattle is restricted to help preserve the pasture and the groundwater below it. At any given time, there are 500 cows involved in daily milk production. The dairy produces 20,280 liters (5,500 gallons) of milk per day, or 41 liters (11 gallons) on the average from each cow. The dairy farms use their own well water to feed and wash the cows, as well as for milk production and irrigation. Most of the cow feed comes from grain grown on other farms on the island—the rest is from the farms' pastures.

   Will the existing dairy farms meet your needs, or will you need to increase dairy farming or import milk? How will either of these choices impact the island? Consider your information from the scaling-up activity and plan how you will obtain enough milk.

3. **Material resources: Other materials needed**

   You will also need sources of vinegar, ammonia, and other supplies. Consider the properties of these materials as you decide how you will obtain, transport, and store them.

4. **Human resources**

   Workers are needed to run the plant. How many workers do you expect to need? What kind of skills will they need for the kinds of jobs you will have? Where will they live? Do you have enough people to fill the jobs on the island, or will you have to hire people from the mainland? The production process may be altered so that less labor is required to operate the factory, but this change would double the amount of energy required to run the plant. More oil would have to be imported or more water power used, which would drive up the energy costs.

## Planning a Factory

**5. Energy resources**

You will need energy for power to run the factory machinery, lights, heating system, and other tasks. You must consider whether the current energy supply can meet your needs. If not, then you will have to find a way to obtain energy by expanding existing power plants or finding new energy sources. You must consider the environmental impact of any choice. Oil and gasoline are needed to run cars and trucks and to lubricate machinery. To import oil, you will need to have a transport system—ships, docks, pipelines, trucks—in place. Solar power can meet some of your energy needs but is a more expensive and seasonal energy alternative, except for heating water for the plant.

**6. Water**

The island currently has enough water to meet its needs. The western and central areas use groundwater. In the southeast, most people have their own wells, or they take water from streams or small reservoirs. In planning your factory, you must consider the quantity and quality of the island's water supply.

**7. Land use**

A limited number of sites are suitable for construction of the factory and related facilities on the island. You must take care to come up with a plan that provides for safety and has the least negative effect on the island's environment. The tourism industry is very concerned about which land is used for the factory. The location of the factory will also affect the costs of transportation for workers and materials. Some land use can be decreased by providing for certain needs off the island.

**8. Waste disposal**

How can you treat and/or dispose of wastes? Consider the impact of your waste treatment plan on the environment and on human health and safety.

# 64 Planning a Factory

## Procedure

*Part Two: Environmental Impact Report*

Your environmental impact report should be a balanced presentation of both negative and positive impacts of the factory on the natural and human environment of Pinniped Island. Your report should include at least the items in the outline presented below.

- Your company is encouraged to add additional information to your report to strengthen your presentation.

- In preparing your report, give careful consideration to the background information provided to the Town Council on pages 38–45 of your Student Book.

- Your company will have 10 minutes to present your plan to the council.

- Your plan will be rated according to how well it considers community concerns.

| | |
|---|---|
| **1. Factory proposal** | Describe your factory in detail. Consider the entire life cycle of the glue, from obtaining the materials needed to the packaging and shipment of the product to its final market. Choose and justify the site of your factory. Your plan should describe the proposed physical layout of the factory and the chemical interactions that the factory will use. If you plan to build power plants, roads, docks, or other facilities on the island, you should also indicate the locations of these facilities. |
| **2. Existing environment** | Where will you locate the factory? Describe the environment as it now exists. |

*Planning a Factory*

**3. Factory impact on the natural environment**

Describe how your project will impact the natural environment. Include both positive and negative effects. Be sure to include the following aspects of the natural environment. You may wish to discuss additional ones as well.

    a. surface water

    b. groundwater

    c. vegetation

    d. animal life

**4. Factory impact on the human environment**

Describe how your factory will impact the human environment. The "human environment" refers to any feature of the environment related to human activity, from buildings and cities to economic, social, and political systems. Your report should include the following areas plus any additional areas that you think are important.

    a. water supply

    b. transportation systems

    c. waste systems

    d. community organizations (such as schools, recreational facilities, housing, hospitals)

    e. economic factors

    f. quality of life (such as noise, lighting, air, visual environment, factors that could involve health risks)

**5. Mitigating impacts**

Describe how you plan to mitigate (lessen or compensate for) any negative impacts you described in Sections 3 and 4. Also describe any impacts that cannot be mitigated.

**6. Summary statement**

Write a brief statement describing the major tradeoffs that you encountered in developing your plan. Indicate why you think that your plan is the best overall factory plan possible.

**7. Resources**

Describe the resources that you used in preparing your report.

*Planning a Factory*

## Water Report

### Prepared by Pinniped Island Water and Sanitation Company

Clear Lake and Wells 1 and 2, currently owned by the company, provide the drinking water supply on Pinniped Island. We removed Well 3 from service in 1993, when iron levels increased with a rise in the acidity of the water. We are investigating recent increases in nitrates in Well 1, although they are still well below federal standards. Clear Lake serves as a reservoir for emergency water needs and is used for recreational purposes, including swimming, boating, and fishing. Because of drainage into the lake from unknown sources, the level of THMs has increased by 0.01 ppm over the last year.

Currently, Clear Lake and Wells 1 and 2 are able to supply 180 gallons of water per minute to the water treatment plant in Wright's Ferry. Here, the water is chlorinated and filtered before it is pumped to customers in the service area. The company has 2,320 metered customers, serving more than 8,500 people. The average person uses 27 gallons of water per day on the island. Other people on the island obtain their water from streams, small reservoirs, and private wells.

Your water company currently operates a waste-water treatment plant located to the south of Wright's Ferry. The plant treats waste water (effluent) to a "primary level." The effluent is first placed in large settling tanks to separate out liquids and solids. The remaining liquid effluent is chlorinated and pumped through an underwater pipe one-half mile out into the ocean, where it disperses in deeper water. The solid material is composted, dried, and sold to the island residents for use as soil amendments. The water treatment system handles 175,000 gallons of waste per day and has a capacity of 200,000 gallons per day.

# Planning a Factory

| Federal Standards (ppm) | | Clear Lake | Well 1 | Well 2 | Well 3 |
|---|---|---|---|---|---|
| Chloride | 250.0 | 47.0 | 15.5 | 14.0 | 220.0 |
| pH | 6.5–8.5 | 6.6 | 7.2 | 6.8 | 6.3 |
| Turbidity | 1.0 (NTU) | 0.1 | 0.2 | 0.2 | 1.3 |
| Coliform bacteria | < 5% | 3.5 % | 1% | 0.5% | 0% |
| Total dissolved solids | 500.0 | 120.0 | 87.0 | 85.6 | 520.7 |
| Nitrate | 10.0 | 1.4 | 3.2 | 0.6 | 20.1 |
| Iron | 0.30 | 0.10 | 0.25 | 0.25 | 0.60 |
| Trihalomethanes (THMs) | 0.10 | 0.02 | 0.03 | 0.02 | 0.01 |

Many people on the island use septic tank systems, where their household waste water is collected in large underground tanks. Here, solids settle and bacteria help decompose the mixture. The liquid portion in each tank overflows into a pipe that leads to a group of branching, porous pipes buried underground—the leach field. In the leach field liquid wastes feed into the subsoil and gradually spread out away from the home.

# Planning a Factory

## Energy Summary
### Pinniped Island Energy Future—From the '90s to the 21st Century

Prepared by SR Associates for the Pinniped Island Business Council

### Energy generation

Pinniped Island uses one fuel oil-powered generating plant located on the coast, southwest of Pinniped Ferry. Along with fuel oil, gasoline, propane, and diesel fuel are imported and stored in a "tank farm" near the power plant. Island Fuel and Service Company distributes these fuels to businesses and homes across the island.

### Current energy use

At the right is a pie chart that shows electrical power usage, as well as a graph of the historic demand for power.

The generating facility is capable of producing up to 500,000 kilowatt hours of electrical energy per day. About 14,500 customers currently use the system.

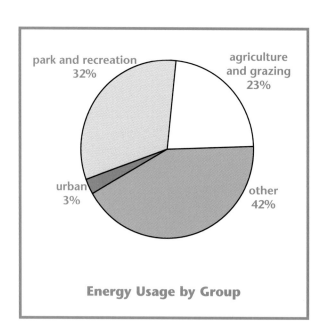

**Energy Usage by Group**

# Planning a Factory

## Future outlook

With increasing fuel prices, the island must consider other alternatives to its present dependency on imported petroleum products. Several alternatives are available: strict conservation measures to reduce electrical demand; the construction of a wind farm at the northern point of the island; the use of solar collectors and cells to generate energy, using large amounts of land to generate solar photovoltaic power (the island has over 200 sunny days per year); and the conversion of biomass from sheep and cattle. The latter would create methane gas that could be burned in the electrical plant to create energy.

If these alternatives are not acceptable, the only way to increase the amount of power generated would be to add a second electricity-generating facility dependent on fossil fuels. This facility would cost $17 million to construct and would require more imported fuels. Another alternative would be to dam some of the island's rivers to create hydroelectric plants.

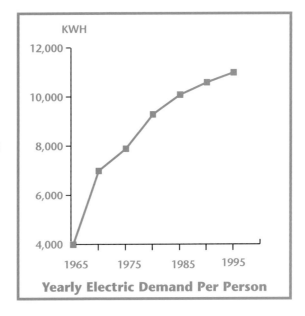

**Yearly Electric Demand Per Person**

*Planning a Factory*

# Land Use Report

**Prepared by Pinniped Island Bureau of Land Management**

*The Pinniped Island Bureau of Land Management manages the island's parks and recreation areas and forests. The BLM also works with island residents to study agricultural land use and other uses of island resources.*

## Public Lands

The BLM manages the island's two regional parks (High Mountain and Clear Lake Parks) and numerous smaller recreation areas. Our newest area is the Buck's Marsh Wildlife Refuge north of Buck's Harbor. This refuge has been set aside to protect species of birds and other wildlife on the eastern shore. Bird populations and other wildlife remain healthy on the island, and we hope to continue our tradition of respect for the island's diverse wildlife.

Plans are under way to protect parts of the forest east of Clear Lake in the future. Other forested areas are important in timber production, which is restricted to meeting island needs. Lumber is not exported from Pinniped Island.

Mining resources on the island are limited. The limestone and marble quarries in the western areas are controlled by the BLM.

*Planning a Factory*

# 64

## Agriculture on Pinniped Island

Nearly 70% of the agricultural land on Pinniped Island is made up of large ranches and farms that are devoted to grazing sheep and cows and/or raising corn and wheat used to feed livestock. The remaining agricultural land consists of a number of small farms that primarily produce food crops.

In planning future agricultural land use on Pinniped Island, it is important to note that a substantial portion of the island (20%) is extremely hilly or mountainous and not suitable for agriculture. The central valley, parts of the area south of Wright's Ferry, and areas of the northern and southwestern peninsulas have suitable land that is not yet devoted to agricultural use. Development of these areas as farmland should be accompanied by careful studies of the impact of farm animals on vegetation and water systems.

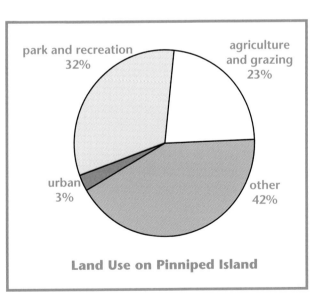

**Land Use on Pinniped Island**

*Activity 64*

# Employment Report
## Pinniped Island Department of Vital Statistics

*The island population has increased steadily since 1960. Rapid increases occurred as a result of the major tourist industry growth period of the '60s. This led to a sharp increase in the birth rate in the late '70s and early '80s.*

## Occupation and Income of Island Residents

| Job | Percent of Population | Median Income |
|---|---|---|
| Fishing | 15% | $17,000 |
| Tourism | 32% | $28,000 |
| Dairy farming | 4% | $26,000 |
| Sheep raising | 4% | $28,000 |
| Other farming | 4% | $24,000 |
| Professionals* | 7% | $35,000 |
| Labor** | 4% | $22,000 |
| Small businesses*** | 12% | $21,500 |
| Other | 8% | $14,000 |
| Unemployed | 10% | unknown |

\* Includes doctors, lawyers, teachers, engineers, and other positions generally requiring a college degree.
\*\* Includes construction workers, transportation workers, etc.
\*\*\* Includes small businesses that serve year-round and summer residents.

*Planning a Factory*

# 64

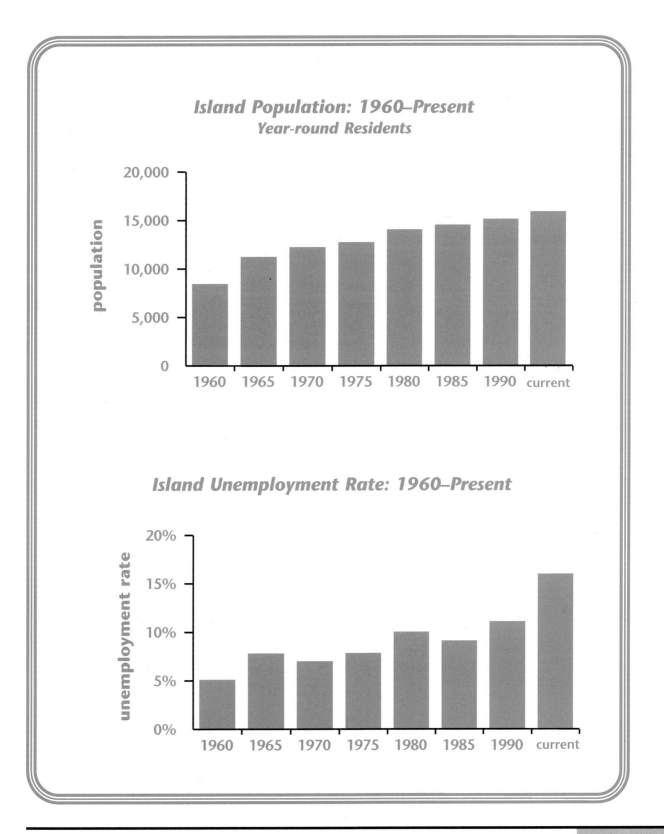

# Activity 65

## An Island Parable

### Introduction

**Easter Island**

The parable of Easter Island describes what can happen when people do not consider all the implications of their actions.

### Challenge

Identify the problems that the settlers of Easter Island faced.

Statues on Easter Island

# An Island Parable

They came in long canoes, riding majestically across the clear blue waters of the Pacific. They came, bringing hundreds of vegetable and fruit plants to cultivate for food.

They came to a wind-swept island full of forests and flowers, covered everywhere with the volcanic debris of the towering brown mountain.

Their plants grew and they prospered. Fields of bananas, yams, and ti (a tropical palmlike tree) covered the rich brown soil. They cut trees to make more canoes to fish the abundant waters and build their homes. Soon, their population grew, and they needed more land to grow food and more wood to shelter their growing numbers against the fierce, damp winds of winter.

Out of the volcanic rocks of the mountain they carved huge humanlike statues, which they hauled down its slopes on wooden sleds to be erected as giant figures on platforms by the edge of the sea, staring out into the vastness of a never-ending ocean.

Over many hundreds of years, the forests were slowly cut down. And as this happened, the newcomers arrived—the ones with "big ears" who came in their canoes. They, too, desired to share the bounty of the island. But they scorned the old ways of the islanders. "Why spend time carving stones when there is so little food to be had here?" they asked.

It was true. The once rich island was now treeless and full of houses—people even lived in rocky caves formed from long lava tubes. The people now numbered nearly ten thousand. There was no longer time to create those majestic statues. Now there was only time to survive and guard one's food and shelter from the newcomers.

The newcomers fought for land and eventually overcame the statue-carving "short ears," whom they had grown to hate. In a final battle, nearly all of the short ears were obliterated from the island. This one-day war ended with all of their bodies being placed in a giant fire in long trenches dug along the side of the dead volcano.

But soon, too soon, the food supply ran short. Eventually, the big ears turned to cannibalism. Skeletons filled the empty caves. The remaining wood was burned to cook food and fend off the bone-chilling winds of winter.

One day, a ship arrived to find fewer than 50 natives, eking out a meager existence on a rock-filled island covered with short grasses and the remains of village dwellings. And yes, there were those mysterious statues, made in a far distant time, looking out to the vast sea.

The name of the island, you ask? It is called Easter Island, found 2,300 miles off the coast of Peru and nearly 1,200 miles from the nearest island in Polynesia.

### Questions

1. What could the settlers of Easter Island have done to prevent the problems they faced?

2. What does this parable tell us about our own relationship to our environment?